Progress in Molecular and Subcellular Biology

Series Editors: W.E.G. Müller (Managing Editor), Ph. Jeanteur, I. Kostovic, Y. Kuchino, A. Macieira-Coelho, R.E. Rhoads

24

Springer

Berlin
Heidelberg
New York
Barcelona
Hongkong
London
Milan
Paris
Singapore
Tokyo

Alvaro Macieira-Coelho (Ed.)

Cell Immortalization

With 13 Figures

 Springer

Professor Dr. A. MACIEIRA-COELHO
INSERM
73 bis, Rue du Meréchal Foch
78000 Versailles
France

ISBN 3-540-65618-9 Springer-Verlag Berlin Heidelberg New York

Library of Congress Cataloging-in-Publication Data.
Cell immortalization/Alvaro Macieira-Coelho (ed.). p. cm. – (Progress in molecular and subcellular biology; 24)
Includes bibliographical references and index.
ISBN 3-540-65618-9
1. Cell proliferation. 2. Apoptosis. I. Macieira-Coelho, Alvaro, 1932 – II. Series. QH506.P76
no. 24 [QH604.7] 572.8 s [571.8'4] – DC21

Production: Pro Edit GmbH, D-69126 Heidelberg, Germany
Cover design: Meta Design, Berlin, Germany
Typesetting: Mitterweger Werksatz GmbH, D-68723 Plankstadt, Germany
SPIN: 10675035   39/3136   5 4 3 2 1 0 – Printed on acid-free paper

Preface

The problem of the long-term proliferation of cells is a seminal one. It has always been a hot subject in biology, a source of far-reaching hypotheses, even more so now when explanations for the mechanisms of cell proliferative mortality or immortality seem within our reach.

A question which is still debated is whether an infinite division potential can be a normal trait or is always the result of modifications leading to abnormal cell growth and escape from homeostasis. In general, investigators have been advocates of one of the two extremes, universal limited or unlimited normal proliferative potential.

Since the long-term proliferative potential of cells concerns regulation of development, regeneration of tissues, and homeostatic control of cell growth, in brief survival of living organisms, and since the regulation of these processes is so different along the evolutionary scale, it is not surprising that there does not seem to be any universal trait.

The question of whether cells are endowed with finite or infinite proliferative phenotypes has to be seen using the perspective of comparative biology. Phenotype depends on kingdom, phyla, order, species and cell type. The two phenotypes, mortal or immortal, have unravelled through evolution. The preservation of an immortal phenotype seems to have given, in some instances, an advantage for survival, in terms of the organism (e.g. unicellular organisms or stem cells in metazoans) and also of the biosphere (e.g. the plant kingdom). These traits, however, cannot be seen only from the point of view of teleonomy. The maintenance of cell mortality or immortality is a phenomenon that evolved with other characteristics of the organism, related to phylogenic, ontogenic and pathogenic characteristics which are described in this volume. From an evolutionary point of view, it must be related to the diversification of cell functions in increasingly complex organisms.

In cells endowed with a limited cycle of replications, the genome can suffer modifications in vitro, leading to unlimited proliferation. The probability of this event occurring varies with the species.

Some simple hypotheses have been proposed to explain the finite proliferative potential, such as an evolutionary mechanism for avoiding abnormal growth, in other words, cancer. Plants, however, are not systematically eliminated by tumors, although their cells are endowed with an immortal phenotype. Somatic cells from some species (e.g. fishes) seem to have an immortal phenotype but these organisms do not develop more cancers than other species. Mammalian stem cells seem to have an infinite division potential but their recruitment is contained according to the needs of the organism. Cells from some mammals transform and immortalize with a high frequency in vitro, however, the organisms in vivo does not produce tumors with increased frequency. Moreover, susceptibility to transformation varies during proliferative aging depending on the carcinogen. Finally, immortalization is often a late phenomenon in the evolution of tumors. Hence, there are no simple explanations and the implications of the presence of a mortal or an immortal phenotype are much more complex; some implications that are looming are described in this volume. Further understanding of the mechanisms involved will reveal fundamental properties of living organisms. The search for the truth has been rewarding; the study of cell immortalization has helped us identify regulatory mechanisms at the molecular level whose relevance for the problem is not yet clear but which are fundamental in molecular and cellular biology. It has also helped to understand several aspects of the biology of cancer.

We felt it was time to make an appraisal and present different views of the subject, emphasizing the certitudes and the ambiguities surrounding the phenomenon.

Versailles, July 1999   Alvaro Macieira-Coelho

Contents

A Theory on Cellular Aging and Cell Immortalization

J. W. I. M. Simons[1]

1
Introduction

Since the early days of tissue culture, the life span of cells has been a mystery: how does the inexorability of death show up in cultured cells? For Alexis Carrel and others, this appeared to be more than a scientific question, as if there was a lesson to be learned about life itself (Witkowsky 1980).

After it became clear that the original statements of Carrel on the unlimited life span of somatic cells were not rooted on firm ground (Hayflick 1989), the senescence and mortality of normal cells in culture did become almost a dogma. Not that the question was solved: it was felt that in vivo the situation might be different, as there is ample reason to believe that there must be stem cells with extraordinarily long life spans (Cameron 1972, Potten and Loeffler 1990). Nowadays the mystery has faded. The ubiquitous observation that, as a rule, tumor cells and cells from established cell lines, thus inferior cells, are immortal has not raised much surprise, which is rather astonishing considering the vast accumulation of damage and disorder in these cells.

Recently, the discovery that the introduction of telomerase activity in normal human cells causes an extension of the replicative life span (Bodnar et al. 1998, Vaziri and Benchimol 1998) has renewed the old quest: has the fountain of youth been detected? This finding indeed suggests that, in principle, immortality might be the rule and that cellular senescence is a cellular program like apoptosis or mitosis that can be switched on or off. All the theories then on the accumulation of damage during cellular senescence seem superfluous or are reduced to the question of whether they can cause telomere shortening (Zglinicki and Saretzki 1997).

However important the contribution of telomere-telomerase research will be the coming years, it can be stated beforehand that in a strict sense the telomere-telomerase hypothesis will not solve the phenomenon of senescence of normal cells, nor explain the appearance of immortality in tumor cells. In a strict sense, the telomere-telomerase hypothesis links the initia-

---

[1] Department of Radiation Genetics and Chemical Mutagenesis, MGC, Leiden University, Wassernaarseweg 72, 2333 AL Leiden, The Netherlands.

Progress in Molecular and Subcellular Biology, Vol. 24
A. Macieira-Coelho
© Springer-Verlag Berlin Heidelberg 1999

tion of cellular senescence to the critical shortening of telomeric DNA. Although this appears perfectly true, there are many phenomena in senescence which do not relate to short telomeres. First, there is the phenomenon of fast senescence which occurs so rapidly that shortening of telomeres will be minimal and thus cannot be responsible for senescence. For example, primary mouse epidermal fibroblasts start to senesce immediately after seeding (J.W.I.M. Simons unpubl. data); human lens epithelial cells senesce within three passages (Andley et al. 1994); immortal Li-Fraumeni fibroblasts senesced within 6 days after treatment with 5-aza-2'-deoxycytidine (Vogt et al. 1998); conditionally immortalized cells demonstrate the senescent phenotype soon after suppression of the SV40 large T antigen (Jiang and Ringertz 1997) and senescence can be quickly induced in NIH 3T3 cells by transfection of mot-1 cDNA (Wadhwa et al. 1993a). Secondly, senescence can occur in the absence of telomere shortening. For example, human uro-epithelial cells in culture have a limited life span despite the presence of telomerase and the absence of telomere shortening (Belair et al. 1997); also, human oral keratinocytes have a limited life span without shortening of telomeres (Kang et al. 1998). Thirdly, telomerase does not seem to have an active role in the appearance of immortality. For example, immortalization of human mammary epithelial cells can take place in the absence of telomerase activity and with shortening of telomers (Stampfer et al. 1997), which is similar to observations made in mouse mammary carcinogenesis (Jiang et al. 1997). Furthermore, a definitive role of telomerase activity is difficult to reconcile with the multistep nature of immortalization (Bols et al. 1991; Namba et al. 1993).

Therefore, the telomere-telomerase issue will be just another, albeit important, link in the cellular mechanisms involved in senescence and unlimited life span. Indications for relations of telomerase with proliferation rate (Belair et al. 1997) and cell cycle (Harley and Sherwood 1997) have already been found and possibly many more will turn out to exist.

This chapter tries to provide a framework for cellular senescence and immortalization in which all the existing facts and theories can find a place. Although highly speculative, checking of the validity of predictions made with respect to aging and immortalization will teach us about life.

2
Preponderances of a General Theory on Cellular Aging and Immortalization

In a strict sense, it has not been proven that aging at the cellular level contributes to or is responsible for the aging of individuals (Barrett et al. 1993). It can, however, be expected that animal senescence will depend on cellular aging, as cellular aging will affect the functioning of tissues, and with it the interactions of tissues (Van Gansen and Van Lerberghe 1988). Therefore, it appears reasonable to suppose that many aging changes in an individual have a cellular basis (Cristofalo et al. 1994).

Before trying to describe the phenomena that a theory on cellular aging would have to cover, a couple of important questions have to be answered. Firstly: is cellular aging a complex phenomenon? Generally it is stated that aging is a very complex process (Kirkwood and Franceschi 1992, Cristofalo 1996, McClearn 1997). Certainly the manifestations of aging at all levels of the organism are very complex as animals are very complex systems, but our knowledge on the cellular aging process appears so limited that it seems impossible to make a solid statement on complexity or simplicity of the cellular aging process. A second basic question is whether aging should be considered as deleterious. At first sight this might seem a superfluous question, used as we are to aging as a threat. Nevertheless, it is quite possible that many manifestations of aging are the result of adequate solutions to deleterious influences. Such adaptations might happen at all levels of the organism and also at the cellular level. Therefore, before even studying aging, it would be wise to keep in mind that aging might be a successful mechanism for coping with deleterious events.

There are a number of phenomena any theory on aging would have to explain. One is that aging can be manipulated to a great extent as is shown by the effect of caloric restriction on rodents and the effect of some mutations on the life span of nematodes, yeast and *Drosophila* (Rose and Nusbaum 1994). Also, the life span of cultured cells can be manipulated to a great extent (Macieira-Coelho 1966, Loo et al. 1987, Macieira-Coelho 1996). For the genus *Hydras*, aging even seems absent due to a stock of immortal embryonic cells (Van Gansen and Van Lerberghe 1988).

The existence of normal immortal cells or cells with a very long life span constitutes a second phenomenon that has to be explained. The life span of some cell types in vivo are estimated to be a 1000 population doublings (Moog 1977) which feeds the notion that cellular aging is a characteristic of differentiated tissues, but need not be a necessity for stem cells (Trosko and Chang 1989).

A third important phenomenon in aging seems to be heterogeneity in the aging process. The subsystems of an organism seem to have different biological ages (McClearn 1997), clonal populations have different life spans (Harley and Goldstein 1980) and all tissues and cell types appear to differ in the aging process (Cristofalo et al. 1994). Within the cell, hundreds of variables are involved in aging (Hayflick 1989), which display increments and decrements (Hayflick 1984) and the number of genes involved in aging could even be several thousands (Martin 1980). Therefore heterogeneity of aging processes appears to prevail.

Moreover, an explanation is needed as to why aging is irreversible, both at the level of the individual and at the level of the cell, and whether there are grounds for the strong impression that some changes in aging proceed by saltation rather than continuously (McClearn 1997).

A fifth phenomenon to be explained by an aging theory is the existence of immortalized abnormal cells. How can they exist? This question has not been raised often, probably because we are so familiar with established cell lines, but it is still unknown how an inferior cell accomplishes what normal cells cannot. As explanation it has been put forward that immortal cells would grow more vigorously, and so outgrow cells with defects (Kirkwood 1991). But many immortal cells do not grow vigorously. Another explanation supposes that in immortal cells special maintenance mechanisms are derepressed (Kirkwood 1991). Surprisingly little research appears to have been carried out on the comparison of mortal and immortal cells with respect to the supposed driving forces of senescence such as the condition of the mitochondria, the presence of aberrant proteins, the protein synthesis rate and turnover, anti-oxidant enzymes, post-translational modifications and the accumulation of waste. Some cancer cell lines seem deficient in the anti-oxidant enzymes superoxide dismutase (Dionosi et al. 1975, Bozzi et al. 1976) and catalase (Bellisola et al. 1987). Further, we came across a four-fold increased rate of mistranslation in immortal cells compared with their normal counterparts (Harley et al. 1980). In addition, it is a fact that cells can still be immortal when they are crippled by severe DNA repair deficiencies. Therefore, under the assumption that the maintenance mechanisms of immortal cells are not superior to those in mortal cells, an important conclusion has to be drawn: conditions can exist in which the occurrence and accumulation of errors does not lead to cellular aging. As disruptive events appear unavoidable in living cells, this would imply that a distinction has to be made between disruptive events and the aging process. The latter clearly is a characteristic of a normal cell and it breaks down during immortalization. Apparently this breakdown is not completely irreversible because, after introduction of senescence genes, immortal cells can become committed to senescence again. Therefore, a theory on aging would have to describe the essence of the process which breaks down during immortalization. Moreover, the theory would have to explain why immortalization can take place as a multistep event (Bols et al. 1991, Namba et al. 1993) and why, after treatment of cells with carcinogens, the progeny of each treated cell has an enhanced chance for immortalization (Bols et al. 1992, Trott et al. 1995).

3
The Aging/Immortalization Hypothesis

The numbers of theories on cellular aging are manifold. However, it seems that they can be classified into two main groups (Martin 1980): one group of theories based on the accumulation of damage and a second group based on a built-in program of cellular senescence. Thus, both classes of theories assume the existence of a tight control mechanism for cellular homeostasis that is or progressively damaged in the course of time or develops according to a planned scheme into a senescent mode. Thus it is assumed that living systems can and do possess the possibility for a completely balanced mechanism of cellular homeostasis. As far as we know, no evidence exists to support this assumption. Therefore, the hypothesis described here challenges this very assumption. Instead, it is assumed, as put forward by others (Bortz 1986, Macieira-Coelho 1990, Toussaint et al. 1991, McClearn 1997), that systems are not in equilibrium: despite all complex regulation, systems will always progressively change, deteriorate and finally collapse. The driving force for this change is formed by all normal fluctuations, which affect the processes within the cell, such as fluctuations in temperature, ionic composition, concentration of chemicals, pH etc etc. Also, errors caused by radicals or chemical instability will contribute. Inevitable progressive dysregulation will be the result at all levels of structure, organization and functioning of the cell. Thus this dysreglation is considered to originate from normal variation in regulated processes and not only from „wear and tear". The normal variation is supposed to be an inherent instability which increases with time as is expected for non-linear open systems (Toussaint et al. 1991), a process that leads to progressive dysfunction of cells and that is operating at every level of organization in the organism. Thus, in principle, aging is a simple and universal process, although the actual aging history will differ from cell to cell and from individual to individual. Therefore, in our view, aging should not be considered as enigmatic. Also, during aging, there is probably no specific main structure or mechanism that becomes predominantly altered, because that would mean that this particular aging process would determine the rate of aging in all members of a species and thus would be rate-limiting. It is supposed that natural selection would prevent such rate-limiting aging processes.

The second pillar of our hypothesis is that there will be two types of aging processes within the living cell. This idea is demonstrated with the following model of a very simple regulatory mechanism. Two genes, A and B, interact: the product of gene A causes the transcription of gene B and the product of gene B causes the transcription of gene A. Thus these two genes regulate the transcription of each other. If we imagine that this is the only process that takes place in a cell, it is clear that there will be no homeostasis. There will only be two possibilities: or a progressive accumulation of both products or the concentration of both products will become less and less with time. Both processes are considered as true but antagonistic aging processes and it is hypothesized that both processes take place in cells during aging. This model of regulation is an oversimplification of the real situation in the cell in which there are abundant feedback loops and each process might influence all other processes. Nevertheless, in our view, similar gene product increases or decreases are bound to also occur in the complicated systems of gene interactions, and thus also in the absence of damage to the participating genes. Therefore, it is hypothesized that there exist two modes of aging within cells. One aging mode will be characterized by increasing numbers of interactions between gene products, and this progressive aging process is supposed to take place during differentiation and senescence. The other aging mode will be characterized by a decreasing number of interactions of gene products, and this regressive aging process could take place during dedifferentiation and regeneration. Therefore, for convenience, we will refer to these two aging processes as the differentiating and dedifferentiating modes of aging respectively.

Our third assumption is that within a normal cell the consequences of all disturbing fluctuations are channeled into only one predominant, co-ordinate mode of aging with as result that a cell will be either in a differentiating mode or in a dedifferentiating mode of aging. Ultimately, if programmed cell death would not occur, both aging modes are thought to proceed until the collapse of the regulatory systems of the cell.

4
Consequences of the Aging/Immortalization theory

In view of this hypothesis, the questions as to whether aging is a complex process, and whether aging should be considered as deleterious, can be answered. In principle, cellular aging can be considered as a simple process of progressive unavoidable increase in variation and with it dysregulation. Although the final outcome of aging will be the ending of the existence of the organism, the process itself should not be considered deleterious, as there is no distinction possible between aging and living. Therefore, cellular aging, as the channeling of the disturbing events, can be considered as the mechanism which determines the rate of increase in dysregulation. Differences between species in the rate of aging will depend on differences in the efficiency of the cellular regulatory mechanisms.

Although, in principle, a simple mechanism, the actual description of an aging process will probably be as complex as the cell itself, because the number of variables is endless.

Another important consequence of the hypothesis is that, as both aging modes are antagonistic, cellular aging is not a necessity if there is a controlling factor that can switch the cell from the differentiating mode to the dedifferentiating mode and vice versa. In such a case, collapse of the system will not be reached and it is hypothesized that such a mechanism is responsible for the very long life span of stem cells or even their immortality. The immortality of established cell lines and tumor cells is supposed to be due to the breakdown of the unidirectional mode of aging of normal cells so that within one cell the differentiating and dedifferentiating modes of aging are present simultaneously in a disorganized fashion so that just by chance unidirectional aging and thus mortality is prevented. This means that in immortalization the essence of the normal cell has been broken down. Although, theoretically, this could happen in one step, it is understandable that the breakdown of the unidirectional mode of aging could be a long process in which multiple steps can be distinguished before complete immortalization has taken place. This would also explain why some tumor cells can still have a limited life span (Mahacek et al, 1993). The general observation that freshly immortalized cells often require careful handling in order to obtain established cell lines (Wadhwa et al. 1991) also fits this picture. It is also easily understood that insult due to carcinogenic treatment can disturb the co-ordinated aging process in all treated cells irreversibly, causing a greater chance of immortalization in the progeny of each treated cell. This would also explain the long latencies with which tumors can arise at sites of exposure to carcinogens. In addition, differences in immortalization frequencies between cells of different species and different tissues could well be due to differences in efficiencies in maintaining a unidirectional mode of aging within the cell.

The theory predicts that there will be three ways to manipulate the rate of aging. The first is to diminish the number of disturbing fluctuations itself e.g. by lowering the temperature or possibly by caloric restriction. The second is by increasing the efficiency of the cellular regulatory mechanisms; such an increase might have taken place during evolution and be responsible for the differences in life span between cells of different species. The third way relates to the replenishment of tissues by stem cells: use of stem cells with a long life span and an efficient mechanism for tissue renewal; the latter might have been efficiently achieved by *Hydras* species.

As the aging process is driven by chance events, it is to be expected that the actual aging process will not be exactly the same in different individuals and that it will also differ from tissue to tissue and even from cell to cell. This would explain differences in biological ages between the subsystems of an organism.

At first sight the apparent irreversibility of aging seems at variance with the supposed antagonistic effects of the two aging modes. This contradiction is resolved when it is taken into account that feedback mechanisms will try to maintain biological processes within certain limits of variation. This process will take place up to the point when maintenance of a certain level of accuracy becomes problematic and major alterations in regulation have to take place. At that point, reversion of such a major alteration might be less reversible or irreversible, while the changes occurring between major alterations might be fully reversible. The principles of thermodynamics also predict that cells will go from one steady state to another (Toussaint et al .1991). In addition, these considerations are in agreement with the impression that aging can proceed by saltation rather than by continuous change.

Therefore, it is indicated that this hypothesis could lead to an understanding of the basic principle of cellular aging and immortalization. Moreover, it is felt that this theory can accommodate most of the available aging theories.

5
Further Speculations:
Interpretation of Biological Phenomena and Predictions

5.1
Two Types of Cellular Organization

If two antagonistic modes of aging exist, and if, as supposed, both modes are compatible with cell proliferation, two types of cellular organization should be identifiable in cell populations as well as a switching mechanism between these two types. Some evidence for this can be found in the literature.

The differentiating mode of aging would be predominantly present in all the cultured cell strains that undergo cellular senescence, while the dedifferentiating mode of aging would frequently occur in stem cells. In vitro representatives of the latter type of growth would be cultures of embryonic stem cells (ES cells) which possibly can be grown indefinitely, as they grow for more than 700 doublings (Niida et al. 1998) and can be considered as normal immortal cells (Suda et al. 1987, Thomson and Marshall 1998). Possibly most stem cells are immortal, as is indicated for the stem cells of the germ line which do not have telomere shortening (Allsop et al. 1992). For hematopoietic stem cells, evidence is accumulating that they do age (Chiu et al. 1996, Van Zant et al. 1997).

The existence of a switching mechanism between the two organization types is indicated from studies on tumor promotion and wound healing. Tumor promoters can cause inflammatory reactions which constitute a profound switch in cell functions (Grinnell 1990, Marks and Fürstenberger 1993). Such changes enable cells to respond to stresses and can lead to an increase in the life span of the cells and to predisposition to immortalization (Steele et al. 1978, Okeda et al. 1990, Bols et al. 1991, 1992, 1993). Reviewing the literature on cellular senescence, immortalization and stress response might uncover possibilities for testing the hypothesis presented here.

5.2
Cellular Aging and Senescence

It has often been questioned whether aging *in vitro* has anything to do with aging *in vivo* (Cristofalo 1996). According to our theory, aging *in vitro* is taking place because of all the variation in cellular processes. Therefore, aging in vitro is a true aging process although it is bound to differ from the aging *in vivo*. Thus the elucidation of aging *in vitro* is relevant for the understanding of organismic aging.

In general, cellular senescence is characterized by
(1) loss of proliferative activity (Macieira-Coelho and Azzarone 1982, Rubin 1997) which corresponds with higher expression of CDK-inhibitors (Wong and Riabowol 1996);

(2) deregulation of gene activity, which corresponds with repression of transcription factors (Sierra et al. 1989, Adler et al . 1996, Dimri et al. 1996);
(3) progressive impairment in adapting to environmental challenges (Adler et al. 1996; Lee et al. 1996), which corresponds with a reduced production of HSPs (heat shock proteins) upon stress (Liu et al. 1989; Choi et al. 1990; Lee et al. 1996);
(4) decreased potential for elimination of dysfunctional cells by apoptosis (Dimri et al. 1996; Warner 1997);
(5) breakdown of the intracellular machinery responsible for protein degradation (Sierra et al. 1989);
(6) increase in heterogeneity between cells (Macieira-Coelho 1995); and
(7) increase in cell size (Simons and van den Broek 1970, 1975; Morocutti et al. 1997).

Although increase in cell size might be connected with disturbed cell volume regulation (Schmidt and Schibler 1995; Lang et al. 1998) it will also be connected with increased protein synthesis (Halle et al. 1997), with defective protein degradation (Sierra et al. 1989) and may even be connected with ectopic expression (Wareham et al. 1987). Therefore, the increase in cell size reflects an accumulation of gene products during differentiative aging.

Although all the above mentioned characteristics suggest that senescence is connected just with deterioration of cell functions, the senescent phenotype of cells appears primarily to result from a well-regulated mechanism. This follows, among others, from the possibility of inducing senescence or senescent-like features in normal dipoid cells by e.g. ceramide (Venable et al. 1995), sodium butyrate (Xiao et al. 1997), activation of p53 (Bond et al. 1996) and inhibition of PI-3 kinase (Tresini et al. 1998). The normal trigger for inducing the senescence process seems to be the shortening of telomeres (Zglinicki and Saretzki 1997), possibly by the silencing of genes in the subtelomeric region (Kim et al. 1996). Some conditions appear to cause a bypass of this senescence programme, for instance treatment with SV40 or inactivation of P21 (Brown et al. 1997), which increases the life span of the cells considerably. Instead of entering senescence, such cells ultimately enter a crisis that can be connected with genetic instability and the cells die or immortalize. However, immortalization of human cells during crisis is still a rare phenomenon. Therefore, it seems that two types of extension of the life span could be distinguished, one caused by abrogation of the senescence program but with continuation of differentiating aging, the other by switching from the differentiating mode of aging to the dedifferentiating mode. The latter possibly could have taken place in the extension of life span observed after treatment with TPA (Bols et al. 1993), treatment with growth factors (Watanabe et al. 1997) or treatment with carcinogens (Stampfer and Bartley 1985; Chang 1986). Abrogation of the senescence programme might explain the inactivation of the so-called M1 mortality mechanism (Shay and Wright 1989). Thus, the senescence programme may function in the cell as the channel to minimizing the consequences of disturbing fluctuations.

5.3
Cellular Immortality

In agreement with the aging/immortalization hypothesis, immortal cells can again become committed to senescence, for instance after transfection of specific genes (Shibanuma et al. 1997, Uhrbom et al. 1997, Ehrenstein 1998), after treatment with sodium butyrate (Xiao et al. 1997) or perhaps even after prolonged incubation under growth constraints (Chow and Rubin 1996). This could imply that in these cells the unidirectional mode of aging becomes restored.

After introduction of senescence in immortal cells by transfection of senescence genes, immortal variants usually re-appear, probably by loss of the correcting gene. However, it has also been observed that after introduction of senescence into a tumor cell by reexpression of the retinoblastoma protein, subsequent abolishment of this expression did not restore immortality, but led to a kind of crisis similar to crises induced by the SV40 large T-antigen (Xu et al. 1997). This observation can be explained by assuming that a perpetuating unidirectional mode of aging was restored in these cells.

As stated above, ES cells are supposed to be characterized by the dedifferentiating mode of aging. In agreement with our hypothesis is the extremely small size of these cells which indicates that the number of gene products is minimal. Normally, ES cells do not show aging phenomena (Suda et al. 1987). Therefore, in order to know whether such cells can have the usual senescence programme, their telomerase activity would have to be abolished. Recently it has been shown that telomerase deficient ES cells acquire a normal rate of telomere shortening but nevertheless go on dividing at a normal rate for 300 divisions and then still go on dividing at a slower rate up to 480 divisions (Niida et al. 1998). This figure differs markedly from the life span of about 20–30 divisions demonstrated by other normal mouse cells (Kaul et al. 1994). Therefore, in agreement with our theory, the aging process in ES cells seems to be quite different from that observed in normal senescing cells, which supports our theory that the aging process in these cells is different from that observed in differentiated cells.

5.4
The Switching Mechanism Between the Two Modes of Aging

The alterations cells undergo in response to stress and tumor promoters might represent a switch between the two modes of aging because tumor promoters can expand the life span of cells (Umeda and Enaka 1981, Deamond and Bruce 1991, Bols et al. 1993). Likewise, injury to the worm *Paranais litoralis* leads to multiplication of stem cells and dedifferentiation of other cells, which appears to extend the life span of this organism (Martinez 1996).

Stress responses are very heterogeneous and depend on cell type and type of stress; e.g. HSP (heat shock protein) expression can be enhanced in inflamed tissue (Handley et al. 1996, Shinoda and Huang 1996) but some inflammatory reactions can be inhibited by stress (Gromgowski et al. 1989, Cahill et al. 1997). Therefore the question to be answered is whether there exists a stress response which is connected with proliferation and with tumorigenesis and which has a constitutive expression in stem cells.

A connection between HSPs and tumorigenesis is indicated by the upregulation of HSP expression in certain tumor types (Schiaffonati et al. 1991, Ferrarini et al. 1992, Yufu et al. 1992, Oesterreich et al. 1993) and because there are several examples of colocalization of HSPs with oncoproteins (Gilson 1993). Moreover, transforming proteins often stimulate expression of HSPs (Liu et al. 1989). Upregulation of HSP expression can even be present in benign lesions, dysplastic tissue (Sugerman et al. 1995) and hyperplasia (Dinda et al. 1998). Moreover, some tumor promoters are capable of enhancing HSP expression (Horn et al. 1994, Holmberg et al. 1997) while some anti-promoters inhibit HSP expression (Elia and Santoro 1994; Horn et al. 1994; Nagai et al. 1995; Hansen et al. 1997; Tosi et al. 1997). Further, it is indicated that the whole heatshock response is disturbed in immortal cell lines, as they are much more sensitive to heat (Richter and Issinger 1986) although the macromolecules of immortal cells denature at slightly higher temperatures than mortal cells (Kaul et al. 1994). Therefore, HSPs appear connected with tumorigenesis.

HSPs are proteins usually expressed at a basal level, and their synthesis is greatly increased as a response to stress (Tanguay 1983; Gilson 1993; Klemenz et al. 1993; Tanguay et al. 1993; Trautinger et al. 1993; Hatayama et al. 1994). As there are several classes of HSPs, each probably with families of genes, and as they have specific cellular localizations, they are presumably involved in all cellular functions such as proliferation, differentiation and apoptosis and may even be constituents of the cytoskeleton (Tanguay 1983). Some, often smaller, HSPs have been found to lower proliferation (Knauf et al. 1992; Kindas-Mugge et al. 1996; Albertazzi et al. 1998; Kindas-Mugge et al. 1998), sometimes in connection with differentiation (Devaja et al. 1997), while some other, often larger HSPs, have been found to correlate with or to enhance proliferation (Pechan 1991; Yano et al. 1996; Zhu et al. 1996; Vargas-Roig et al. 1997). Also, there are indications that HSPs can participate in cell cycle control (van Dongen and van Wijk 1988; Khandjian 1995; Galea-Lauri et al. 1996). Therefore HSPs appear connected with proliferation.

If expression of HSPs is a characteristic of stem cells, resistance to apoptosis might be a further characteristic of these cells. Heat shock induced cell death is usually not apoptotic in nature (Roti Roti et al. 1992), on the contrary, heat shock has been found to induce resistance to apoptosis (Mailhos et al. 1993; Samali and Cotter 1996). Both HSP70 (Wei et al. 1994; Jamora et al. 1996; Mosser et al. 1997) as HSP27 (Mehlen et al. 1997) have been correlated with this resistance. Sometimes, however, expression of HSP70 and apoptosis occur simultaneously (Filippovich et al. 1994; Chant et al. 1996). An important factor for the effectiveness of HSPs in preventing apoptosis could be the accumulation of the protein in the nucleus rather than its presence in the whole cell (Watanabe et al. 1995).

There are indications that the level of expression and inducibility of HSPs in stem cells could well be different from that observed in mortal cells. Mouse ES cells are the only mammalian cells known to have a constitutive HSE-binding activity (heat shock element) at normal physiological temperatures (Murphy et al. 1994). Moreover, HSPs are suspected to play important roles in embryogenesis. During embryogenesis of *Xenopus* there is constitutive expression of many HSPs and an inability to induce HSPs by heat shock until after the midblastula stage (Heikilla et al. 1997). In *Drosophila,* HSPs are not induced before the blastoderm stage (Arrigo and Tanguay 1991). In mouse there is high HSP expression during the pre-implantation phase (Loones et al. 1997) and preovulatory oocytes, which lack a heat shock response, have constitutive HSP-like proteins (Curci et al. 1987). Therefore, it is likely that stem cells will be characterized by their own pattern of HSP expression.

In conclusion, this literature overview indicates that existing data could support the existence of two basic modes of cellular aging and a switching mechanism between them. If so, the question remains as to why the dedifferentiating mode of aging would be involved in tumorigenesis and predispose to immortalization, the more so because embryonic cells are known to be resistant to tumorigenesis (Kondo 1983; Einhorn 1991). This question has been specifically addressed in a previous paper in which it is argued that a risk for immortalization arises when differentiating cells dedifferentiate and acquire stem cell-like properties (Simons, 1999). In such pseudo-stem cells, the switch from the differentiating mode to the dedifferentiating mode could lead to instability in regulatory mechanisms because the differentiating mode would already have led to less reversible or irreversible steps.

5.5
HSPs and Immortalization

That HSPs can be actively involved in the process of immortalization follows from the observations on mortalin. Mortalin belongs to the HSP70 family (Webster et al. 1994; Hadari et al. 1997; Sadekova et al. 1997). It is constitutively expressed in many rat tissues in which the expression pattern agrees with a possible antiproliferative function (Kaul et al. 1997). In normal cells it is expressed in the cytosolic fraction, but upon heat shock a perinuclear localization is found (Kaul et al. 1993). The perinuclear localization has been proven to be a characteristic of immortal cell lines (Kumazaki et al. 1997), as it was observed in three immortal lines from three different mouse strains, in immortal human fibroblasts (Wadhwa et al. 1993b) and in human glioblastoma cells (Takano et al. 1997). The picture is complicated by the finding that the perinuclear mortalins of three immortal mouse lines were not identical: there are two different proteins (mot-1 and mot-2) which differ by two amino acids (Wadhwa et al. 1993c). As mot-1 and mot-2 are genetic in origin, different genes or alleles exist (Wadhwa et al. 1996). The direct involvement of mortalin in cellular senescence appears from the concomitant induction of cytosolic mortalin and senescence in NIH 3T3 cells by transfection of mot-1 cDNA (Wadhwa et al. 1993a). Also, induction of senescence in a human fibroblast line by introduction of chromosome 7 was accompanied by the change from perinuclear to cytosolic mortalin (Nakabayashi et al. 1997) and the same was observed after fusion of immortal NIH 3T3 cells with normal mouse embryonic fibroblasts (Wadhwa et al. 1993d). The direct involvement of mortalin in immortalization is suggested by the finding that perinuclear distribution was found from the earliest step onward in the immortalization process (Kaul et al. 1994), and that in human cell lines four patterns of perinuclear distribution of mortalin can be distinguished that, grosso modo, coincide with the four complementation groups of immortality (Wadhwa et al. 1995).

5.6
Experimental Approaches?

In principle, the aging/immortalization hypothesis appears verifiable.

One approach would be completely theoretical: the development of computer models to investigate whether cellular regulatory mechanisms, exposed to the normal fluctuations within a 3-dimensional system, will indeed undergo progressive deterioration and whether in principle two types of deterioration should be distinguished.

Studies performed on living cells could demonstrate whether similar aging processes take place *in vivo*, as predicted by model systems. However, as single cells would have to be studied ideally, this could be technically very difficult.

Another approach would be research on the existence and characterization of two basic types of cell organization: stem cell-type and limited life span-type. As it is foreseen that normal cells with limited life spans can acquire a stem cell-like growth, which would be connected with extension of the life span and enhanced chance for immortalization, these stem cell-like cells should be identifiable in cell populations. SHE (Syrian hamster embryo) cells could be useful for this purpose when treated with carcinogens, since cell cultures with an enhanced chance for immortalization can be identified before their actual immortalization by their enhanced growth rate and increased cell density (Bols et al. 1993). The stem cell-like cells are possibly characterized by small cell size, activated HSPs, perinuclear localization of mortalin and protection to induction of apoptosis. Also, cell cycle control and the cytoskeleton might be altered (Lin 1997).

In a previous paper (Simons 1999), predictions have been made with respect to the effect of aging on tumorigenesis. The hypothesis indicates that the efficiency with which initiated cells will develop into premalignant lesions will decline with age but that the premalignant lesions which do develop have an enhanced chance for malignancy.

Another consequence of the hypothesis is that culture conditions could be identified that would allow the immortalization of human cells which are presently known to be very resistant to spontaneous immortalization (Chang 1988; McCormick and Maher 1989). In a pilot experiment we cultured human keratinocytes as shedding epithelia (Simons et al. 1996). Mass cultures of mouse keratinocytes, cultured as shedding epithelia, are known to produce foci of immortal cells (Greenhalgh et al. 1989). Keratinocytes from both mouse and man, when cultured as shedding epithelia, are resistant to the induction of apoptosis (unpubl. data). In agreement with the hypothesis, we obtained several clones of human keratinocytes with extended life span in the absence of treatment with a carcinogen.

6
Conclusions

The aging/immortalization hypothesis, based on the assumption that cellular aging results from the progressive increase in variation in a linear open system, can explain the existence of mortal cells, long-lived stem cells and immortal abnormal cells. Two basic modes of aging are supposed to operate, one connected with differentiation (predominantly present in mortal cells) and one connected with dedifferentiation (predominantly present in stem cells). Normal cells are supposed to be characterized by an unidirectional mode of aging while in abnormal immortal cells the unidirectional mode of aging is supposed to have been broken down.

Acknowledgements. The financial support of the Dutch Cancer Foundation, the Association for International Cancer Research and the J. A. Cohen Institute for Radiopathology and Radiation Protection is gratefully acknowledged.

References

Adler V, Dolan LR, Kim J, Pincus M, Barrett JC, Ronai Z (1996) Changes in jun N-terminal kinase activation by stress during aging of cultured normal human fibroblasts. Mol Carcinog 17:8–12
Albertazzi E, Cajone F, Laksmi MS, Sherbet GV (1998) Heat shock modulates the expression of the metastasis associated gene MTS1 and proliferation of murine and human cancer cells. DNA Cell Biol 17:1–7

Allsopp RC, Vaziri H. Patterson C, Goldstein S, Younglai EV, Futcher AB, Greider CW, Harley CB (1992) Telomere length predicts replicative capacity of human fibroblasts. Proc Natl Acad Sci USA 89:10114–10118

Andley UP, Rhim JS, Chylack LT, Fleming TP (1994) Propagation and immortalization of human lens epithelial cells in culture. Invest Ophthalmol Visual Sci 35:3094–3102

Arrigo AP, Tanguay RM (1991) Expression of heat shock proteins during development in *Drosophila*. Results Probl Cell Differ 17:106–119

Barrett JC, Annab LA, Fureal PA (1993) Genetic and molecular basis for cellular senescence. Adv Exp Med Biol 330:27–43

Belair CD, Yeager TR, Lopez PM, Reznikoff CA (1997) Telomerase activity: a biomarker of cell proliferation, not malignant transformation. Proc Natl Acad Sci USA 94:13677–13682

Bellisola G, Casaril M, Gabrielli GB, Corrocher R (1987) Catalase activity in human hepatocellular carcinoma (HHC). Clin Biochem 20:415–417

Bodnar AG, Ouelette M, Frolkis M, Holt SE, Chiu C, Morin GB, Harley CB, Shay JW, Lichtsteiner S, Wright WE (1998) Extension of life span by introduction of telomerase into normal human cells. Science 279:349–352

Bols BLMC, Naaktgeboren JM, Simons JWIM (1991) Immortalization of Syrian hamster embryo cells is in itself a multistep event. Cancer Res 51:1177–1184

Bols BLMC, Naaktgeboren CJM, Lohman PHM, Simons JWIM (1992) Immortalization of carcinogen-treated Syrian hamster cells occurs indirectly via an induced process. Cancer Res 53:2253–2256

Bols BLMC, Gillis KCPM, Naaktgeboren JM, Lohman PHM, Niericker MJ, Simons JWIM (1993) Immortalization of Syrian hamster embryo cells: probabilistic event or deterministic process. Cancer Res: 53:4797–4802

Bond J, Haughton M, Blaydes J, Gire V, Wynford-Thomas D, Wyllie F (1996) Evidence that transcriptional activation by p53 plays a direct role in the induction of cellular senescence. Oncogene 21:2097–2104

Bortz WM (1986) Aging as entropy. Exp Gerontol 21:321–328

Bozzi A, Mavelli A, Finazzi-Agro A, Strom R, Wolf AM, Mondovi B, Rotilio G (1976) Enzyme defense against reactive oxygen derivatives II. Erythrocytes and tumor cells. Mol Cell Biochem 10:11–16

Brown JP, Wei W, Sedivy JM (1997) Bypass of senescence after disruption of p21CIPI/WAF1 gene in normal diploid human fibroblasts. Science 277:831–834

Cahill CM, Lin HS, Price BD, Bruce JL, Calderwood SK (1997) Potential role of heat shock transcription factors in the expression of inflammatory cytokines. Adv Exp Med Biol 400B:625–630

Cameron IL (1972) Minimum number of cell doublings in an epithelial cell population during the life span of the mouse. J Gerontol 27:157–161

Chang SE (1986) In vitro transformation of human epithelial cells. Biophys Acta 823:161–164

Chant ID, Rose PE, Morris AG (1996) Susceptibility of AML cells to in vitro apoptosis correlates with heat shock protein 70 (hsp 70) expression. Br J Haematol 93:898–902

Chiu C, Dragowska W, Kim NW, Vaziri H, YuiJ, Thomas TE, Harley CB, Lansdorp PM (1996) Differential expression of telomerase activity in hematopoietic progenitors from adult human bone marrow. Stem Cells 14:239–248

Choi H, Lin Z, Li ZL, Liu AY (1990) Age-dependent decrease in the heat- -inducible DNA sequence-specific binding activity in human diploid fibroblasts. J Biol Chem 265:18005–18011

Chow M, Rubin H (1996) Irreversibility of cellular aging and neoplastic transformation: a clonal analysis. Proc Natl Acad Sci USA 93:9793–9798

Cristofalo VJ (1996) Ten years later: what have we learned about human aging from studies of cell cultures. Gerontologist 36:737–741

Cristofalo VJ, Gerhard GS, Pignolo RJ (1994) Molecular Biology of aging. Surg Clin North Am 74:1–21

Curci A, Bevilacqua A, Mangia F (1987) Lack of heat-shock response in preovulatory mouse oocytes. Dev Biol 123:154–160

Deamond SF, Bruce SA (1991) Age-related differences in promoter-induced extension of in vitro proliferative life span of Syrian hamster cell cultures. Mech Ageing Dev 60:143–152

Devaja O, King RJ, Papadopoulos A, Raja KS (1997) Heat-shock protein 27 (HSP27) and its role in female reproductive organs. Eur J Gynaecol Oncol 18:16–22

Dimri GP, Testori A, Acosta M, Campisi J (1996) Replicative senescence, aging and growth-regulatory transcription factors. Biol Signals 5:154–162

Dinda AK, Mathur M, Guleria S, Saxena S, Tiwari SC, Dash SC (1998) Heat shock protein (HSP) expression and proliferation of tubular cells in end stage renal disease with and without haemodialysis. Nephrol Dial Transplant 13:99–105

Dionosi O, Galeotti T, Terranova T, Azzi A (1975) Superoxide radicals and hydrogen peroxide formation in mitochondria from normal and neoplastic tissue. Biochim Biophys Acta 403:292–300

Ehrenstein D (1998) Immortality gene discovered. Science 279:177

Einhorn L (1991) Can prenatal irradiation protect the embryo from tumor development? Acta Oncol 30:291–299

Elia G, Santoro MG (1994) Regulation of heat shock protein synthesis by quercitin in human erythroleukemia cells. Biochem J 300:201–209

Ferrarini M, Heltai S, Zocchi MR, Rugarli C (1992) Unusual expression and localization of heat-shock proteins in human tumor cells. Int J Cancer 51:613–619

Filippovich I, Sorokina N, Khanna KK, Lavin MF (1994) Butyrate induced apoptosis in lymphoid cells preceded by transient over-expression of HSP70 mRNA. Biochem Biophys Res Commun 198:257–265

Galea-Lauri J, Latchman DS, Katz DR (1996) The role of the 90-kDa heat shock protein in cell cycle control and differentiation of the monoblastoid cell line U937. Exp Cell Res 226:243–254

Gilson G (1993) Heat shock proteins. Bull Soc Sci Med Grand Duche Luxemb 130:37–43

Greenhalgh DA, WeltyDJ, Strickland JE, Uuspa SH (1989) Spontaneous

Ha-ras gene activation in cultured primary murine keratinocytes: consequences of Ha-ras gene activation in malignant conversion and malignant progression. Mol Carcinog 2:199–207

Grinnell F (1990) The activated keratinocyte: up regulation of cell adhesion and migration during wound healing. J Trauma 30 (suppl) S144–S149

Gromgowski S, Yagi J, Janeway CA (1989) Elevated temperature regulates tumour necrosis factor-mediated immune killing. Eur J Immunol 19:1709–1714

Hadari YR, Haring HU, Zick Y (1997) p75, a member of the heat shock protein family, undergoes tyrosine phosphorylation in response to oxidative stress. J Biol Chem 272:657–662

Halle JP, Müller S, Simm A, Adam G (1997) Copy number, epigenetic state and expression of the rRNA genes in young and senescent rat embryo fibroblasts. Eur J Cell Biol 74:281–288

Handley HH, Yu J, Yu DT, Singh B,Gupta RS, Vaughan JH (1996) Autoantibodies to human heat shock proteins (hsp) 60 may be induced by Escherichia coli groEL. Clin Exp Immunol 103:429–435

Hansen RK, Oesterreich S, Lemieux P, Sarge KD, Fuqua SA (1997) Quercetin inhibits heat shock protein induction but not heat shock factor

DNA-binding in human breast carcinoma cells. Biochem Biophys Res Commun 239:851–856

Harley CB, Sherwood SW (1997) Telomerase, checkpoints and cancer. Cancer Surv 29:263–284

Harley CB, Goldstein S (1980) Retesting the commitment theory of cellular aging. Science 207:191–193

Harley CB, Pollard JW, Chamberlain JW, Stanners CP, Goldstein S (1980) Protein synthesis errors do not increase during aging of cultured human fibroblasts. Proc Natl Acad Sci USA 77:1885–1889

Hatayama T, Nishijama E, Yasuda K (1994) Cellular localization of

high-molecular-mass heat shock proteins in murine cells. Biochem Biophys Res Commun 200:1367–1373

Hayflick L (1984) Intracellular determinants of cell aging. Mech Ageing Dev 28:177–185

Hayflick L (1989) Antecedents of cell aging research. Exp Gerontol 24: 355–365

Heikilla JJ, Chan N, Tam Y, Ali A (1997) Heat shock protein gene expression during *Xenopus* development. Cell Mol Life Sci 53:114–121

Holmberg CI, Leppa S, Eriksson JE, Sistonen L (1997) The phorbol ester 12- -O-tetradecanoylphorbol 13-acetate enhances the heat induced stress response. J Biol Chem 272:6792–6798

Horn S, Cohen R, Gertler A (1994) Regulation of heat-shock protein (hsp70) gene expression by hGH and Il2 in rat Nb2 lymphoma cells. Mol Cell Endocrinol 105:139–146

Jamora C, Dennert G, Lee AS (1996) Inhibition of tumor progression by suppression of stress protein GRP78/BiP induction in fibrosarcoma B/C10ME Proc Natl Acad Sci USA (1996) 93:7690–7694

Jiang WQ, Ringertz N (1997) Altered distribution of the promyelocytic leukemia-associated protein is associated with cellular senescence. Cell Growth Differ 8:513–522

Jiang C, Juo L, Said TK, Thompson H, Medina D (1997) Immortalized mouse mammary cells in vivo do not exhibit increased telomerase activity. Carcinogenesis 18:2085–2091

Kang MK, Guo W, Park NH (1998) Replicative senescence of normal human oral keratinocytes is associated with the loss of telomerase activity without shortening of telomeres. Cell Growth Differ 9:85–95

Kaul SC, Wadhwa R, Komatsu Y, Sugimoto Y, Mitsui Y (1993) On the cytosolic and perinuclear mortalin: an insight by heat shock. Biochem Biophys Res Commun 193:348–355

Kaul SC, Wadhwa R, Sugihara T, Obuchi K, Komatsu Y, Mitsui Y (1994) Identification of genetic events involved in early steps of immortalization of mouse fibroblasts. Biochim Biophys Acta 1201: 389–396

Kaul SC, Matsui M, Takano S, Sugihara T, Mitsui Y, Wadhwa R (1997) Expression analysis of mortalin, a unique member of the Hsp70 family of proteins, in rat tissues. Exp Cell Res 232:56–63

Khandjian EW (1995) Heat treatment induces dephosphorylation of Rb and dissociation of T-antigen/pRb complex during transforming infection with SV40 antigen. Oncogene 10:359–367

Kim S, Villeponteau B, Jazwinski SM (1996) Effect of replicative age on transcriptional silencing near telomeres in *Saccharomyces cerevisiae*. Biochem Biophys Res Commun 219:370–376

Kindas-Mugge I, Herbacek I, Jantschitsch C, Micksche M, Trautinger F (1996) Modification of growth and tumorigenicity in epidermal cell lines by DNA- -mediated gene transfer of Mw 27,000 heat shock protein (hsp27). Cell Growth Differ 7:1167–1174

Kindas-Mugge I, Micksche M, Trautinger (1998) Modification of growth in small heat shock (hsp27) gene transfected breast carcinoma. Anticancer Res 18:413–417

Kirkwood TBL (1991) Genetic basis of limited cell proliferation. Mutation Res 256:323–328

Kirkwood TBL, Franceshi C (1992) Is aging as complex as it would appear? New perspectives in aging research. Ann N Y Acad Sci 663:412–417

Klemenz R, Andres AC, Frohli E, Schafer R, Aoyama A (1993) Expression of the murine small heat shock proteins hsp 25 and alpha B crystallin in the absence of stress. J Cell Biol 120:639–645

Kondo S (1983) Carcinogenesis in relation to the stem-cell-mutation hypothesis. Differentiation 24:1–8

Kumazaki T, Wadhwa R, Kaul SC, Mitsui Y (1997) Expression of endothelin, fibronectin, and mortalin as aging and mortality markers. Exp Gerontol 32:95–103

Lang F, Busch GL, Volkl H (1998) The diversity of volume regulatory mechanisms. Cell Physiol Biochem 8:1–45

Lee YK, Manalo D, Liu AY (1996) Heat shock response, heat shock transcription factor and cell aging. Biol Signals 5:180–191

Lin H (1997) The tao of stem cells in the germline. Annu Rev Genet 31:455–491

Liu AYC, Lin Z, Choi H, Sorhage F, Li B (1989) Attenuated induction of heat shock gene expression in aging diploid fibroblasts. J Biol Chem 264:12037–12045

Loo DT, Fuquay JI, Rowson CL, Barnes DW (1987) Extended culture of mouse embryo cells without senescence; inhibition by serum. Science 236:200–236

Loones MT, Rallu M, Mezger V, Morange M (1997) HSP gene expression and HSF2 in mouse development. Cell Mol Life Sci 53:179–190

Macieira-Coelho A (1966) Action of cortisone on human fibroblasts *in vitro*. EXP 22: 390–391

Macieira-Coelho A (1990) Relevance of *in vitro* studies for *in vivo* aging. Z Gerontol 23:130–132

Macieira-Coelho A (1995) The implications of the 'Hayflick limit' for aging of the organism have been misunderstood by many gerontologists. Gerontology 41:94–97

Macieira-Coelho A (1996) Proliferative cell senescence, transformation and the recombination potential of the genome. Exp Gerontol 31:227–234 Macieira-Coelho A, Azzarone B (1982) Aging of human fibroblasts is a succession of subtle changes in the cell cycle and has a short stage with abrupt events. Exp Cell Res 141:325–332

Mahacek ML, Beer DG, Frank TS, Ethier SP (1993) Finite proliferative lifespan *in vitro* of a human breast cancer cell strain isolated from a metastatic lymph node. Breast Canc Res Treatm 28:267–276

Mailhos C, Howard MK, Latchman DS (1993) Heat shock protects neuronal cells from programmed cell death by apoptosis. Neuroscience 55:621–627

Marks F, Fürstenberger G (1993) Proliferative responses of the skin to external stimuli. Environ Health Perspect (Suppl)5:95–102

Martin GM (1980) Genotropic theories of aging: an overview. Adv Pathobiol 7:5–20

Martinez DE (1996) Rejuvenation of the disposable soma: repeated injury extends life span in an asexual annelid. Exp Gerontol 31:699–704

McClearn GE (1997) Biogerontologic theories. Exp Gerontol 32:3–10

McCormick JJ, Maher VM (1989) Malignant transformation of mammalian cells in culture, including human cells. Environ Mol Mutagen 14(Suppl):105–113

Mehlen P, Mehlen A, Godet J, Arrigo AP (1997) HSP27 as a switch between differentiation and apoptosis in murine embryonic stem cells. J Biol Chem 272:31657–31665

Moog F (1977) The small intestine in old mice: growth ,alkaline phosphatase and disaccharidase activities and deposition of amyloid. Exp Gerontol 12:223–235

Morocutti A, Earle KA, Rodemann HP, Viberti GC (1997) Premature cell ageing and evolution of diabetic nephropathy. Diabetologia 40:244–246

Mosser DD, Caron AW, Bourget K, Denis-Larose C, Massie B (1997) Role of the human heat shock protein hsp70 in protection against stress-induced apoptosis. Mol Cell Biol 17:5317–5327

Murphy SP, Gorzowski JJ, Sarge KD, Phillips B (1994) Characterization of constitutive HSF2 DNA-binding activity in mouse embryonal carcinoma cells. Mol Cell Biol 14:5309–5317

Nagai N, Nakai A, Nagata K (1995) Quercetin suppresses heat shock response by down regulation of HSF1. Biochem Biophys Res Commun 208:1099–1105

Nakabayashi K, Ogata T, Fujii M, Tahara H, Ide T, Wadhwa R, Kaul SC, Mitsui Y, Ayusawa D (1997) Decrease in amplified telomeric sequences and induction of senescence markers by introduction of human chromosomes 7 or its segments in SUSM-1. Exp Cell Res 235:345–353

Namba M, Iijima M, Kondo T, Jahan I, Mihara K (1993) Immortalization of normal human cells is a multistep process and a rate limiting step of neoplastic transformation of the cells. Hum Cell 6:253–259

Niida H, Matsumoto T, Satoh H, Shiwa M, Tokutake Y, Furuichi Y, Shinkai Y (1998) Severe growth defect in mouse cells lacking the telomerase RNA component. Nat Genet 19:203–206

Oesterreich S, Weng CN, Qiu M, Hilsenbeck SG, Osborne CK Fuqua SA (1993) The small heat shock protein hsp27 is correlated with growth and drug resistance in human breast cancer cell lines. Cancer Res 53:4443–4448

Okeda T, Yokpgawa Y, Ueo H, Burty M, Ts'O POP, Bruce SA (1990) Two classes of continuous cell lines established from Syrian hamster 9 day gestation embryos: preneoplastic cells and progenitor cells. Cell Dev Biol 26:1157–1166

Pechan PM (1991) Heat shock proteins and proliferation. FEBS Lett 280:1–4

Potten CS, Loeffler M (1990) Stem cells: attributes, cycles, spirals, pitfalls and uncertainties. Lessons for and from the crypt. Development 110:1001–1020

Richter WW, Issinger OG (1986) Differential heat shock response of primary human cell cultures and established cell lines. Biochem Biophys Res Commun 141:46–52

Rose MR, Nusbaum TJ (1994) Prospects for postponing human aging. FASEB J 8:925–928

Roti Roti JL, Mackey MA, Higashikubo (1992) The effects of heat shock on cell proliferation. Cell Prolif 25:89–99

Rubin H (1997) Cell aging *in vivo* and *in vitro*. Mech Ageing Dev 98:1–35 Samali A, Cotter TG (1996) Heat shock proteins increase resistance to apoptosis. Exp Cell Res 223:163–170

Sadekova S, Lehnert S, Chow TY (1997) Induction of PBP74/mortalin/Grp75, a member of the hsp70 family, by low doses of ionizing radiation: a possible role in induced radioresistance. Int J Radiat Biol 72:653–660

Schmidt EE, Schibler U (1995) Cell size regulation, a mechanism that controls cellular RNA accumulation: consequences on regulation of the ubiquitous transcription factors Oct1 and NF-Y, and the liver enriched transcription factor DBP. J Cell Biol 128:468–483

Schiaffonati L, Pappalardo C, Tacchini L (1991) Expression of the HSP 70 gene family in rat hepatoma cell lines of different growth rates. Exp Cell Res 196:330–336

Shay JW, Wright WE (1989) Quantitation of the frequency of immortalization of normal human fibroblasts by SV40 large T-antigen. Exp Cell Res 184:109–118

Shibanuma M, Mochizuki E, Maniwa R, Mashimo J, Nishiya N, Imai S, Takano T, Oshimura M, Nose K (1997) Induction of senescence-like phenotypes by forced expression of hic-5, which encodes a novel LIM motif protein, in immortalized human fibroblasts. Mol Cell Biol 17:1224–1235

Shinoda H, Huang CC (1996) Heat shock proteins in middle ear cholesteatoma. Otolaryngol Head Neck Surg 114:77–83

Sierra F, Fey GH, Guigoz Y (1989) T-kininogen gene expression is induced during aging. Mol Cell Biol 9:5610–5616

Simons JWIM (1999) Genetic, epigenetic, dysgenetic and non-genetic mechanisms in tumorigenesis. II. Further delineation of the rate limiting step. Anticancer Res (in press)

Simons JWIM, van den Broek C (1970) Comparison of ageing *in vitro* and ageing *in vivo* by means of cell size analysis using a coulter counter. Gerontolgy 16:340–351

Simons JWIM, van den Broek C (1975) Studies on the proliferative capacity of mouse spleen cells in serial transplantation. Adv Exp Med Biol 53:219–233

Simons JWIM, Niericker MJ, van Klaveren P, Verdegaal E (1996) Are hyperproliferative shedding epithelia an in vitro counterpart of promoter induced hyperplasia and do they predispose to immortalization? Proc Am Assoc Cancer Res 37:162

Stampfer MR, Bartley JC (1985) Induction of transformation and continuous cell lines from normal human mammary epithelial cells after exposure to benzo(*a*)pyrene. Proc Natl Acad Sci USA 82:2394–2398

Stampfer MR, Bodnar A, Garbe J, Wong M, Pan A, Villeponteau B, Yaswn P (1997) Gradual phenotypic conversion associated with immortalization of cultured human mammary epithelial cells. Mol Biol Cell 8:2391–2405

Steele VE, Marchok AC, Nettesheim P (1978) Establishment of epithelial cell lines following exposure of cultured tracheal epithelium to 12-O-tetradecanoyl- -phorbol-13-acetate. Cancer Res 38:3563–3565

Suda Y, Suzuki M, Ikawa Y, Aizawa S (1987) Mouse embryonic stem cells exhibit indefinite proliferative potential. J Cell Physiol 133:197–201

Sugerman PB, Savage NW, Xu LJ, Walsh LJ, Seymour GJ (1995) Heat shock protein expression in oral epithelial dysplasia and squamous cell carcinoma. Eur J Cancer B Oral Oncol 31B:63–67

Takano S, Wadhwa R, Yoshii Y, Nose T, Kaul SC, Mitsui Y (1997) Elevated levels of mortalin expression in human brain tumors. Exp Cell Res 237:38–45

Tanguay RM (1983) Genetic regulation during heat shock and function of heat shock proteins: a review. Can J Biochem Cell Biol 61:387–394

Tanguay RM, Wu Y, Khandjian EW (1993) Tissue-specific expression of heat shock proteins of the mouse in the absence of stress. Dev Genet 14:112–118

Thomson JA, Marshall VS (1998) Primate embryonic stem cells. Curr Top Dev Biol 38:133–165

Tosi P, Visani G, Ottaviani E, Gibellini D, Pellacani A, Tura S (1997) Reduction of heat-shock protein-70 after prolonged treatment with retinoids: biological and clinical implications. Am J Hematol 56:143–150

Toussaint O, Raes M, Remacle (1991) Aging as a multistep process characterized by a lowering of entropy production leading the cell to a sequence of defined stages. Mech Ageing Dev 61:45–64

Trautinger F, Trautinger I, Kindas-Mugge L, Metze D, Luger TA (1993) Human keratinocytes *in vivo* and *in vitro* constitutively express the 72-kDa heat shock protein. J Invest Dermatol 101:334–338

Tresini M, Mawal-Dewan M, Cristofalo VJ, Sell C (1998) A phosphatidylinositol 3-kinase inhibitor induces a senescent-like growth arrest in human diploid fibroblasts. Cancer Res 58:1–4

Trosko JE, Chang CC (1989) Stem cell theory of carcinogenesis. Toxicol Lett 49:283–295

Trott DA, Cuthbert AP, Overell RW, Russo I, Newbold RF (1995) Mechanisms involved in the immortalization of mammalian cells by ionizing radiation and chemical carcinogens. Carcinogenesis 16:193–204

Uhrbom L, Nister M, Westermark B (1997) Induction of senescence in human malignant glioma cells by p16INK4A. Oncogene 31:505–514

Umeda M, Enaka E (1981) Some aspects of in vitro carcinogenesis using Syrian hamster cell cultures. Gann Monogr Cancer Res 27:183–194

Van Dongen G , van Wijk R (1988) Evidence for a role of heat-shock proteins in proliferation after heat treatment of synchronized mouse neuroblastoma cells. Radiat Res 113:252–267

Van Gansen P, Van Lerberghe N (1988) Potential and limitations of cultivated fibroblasts in the study of senescence in animals. A review on the murine skin fibroblasts system. Arch Gerontol Geriatr 7:31–74

Van Zant G, de Haan G, Rich IN (1997) Alternatives to stem cell renewal from a developmental viewpoint. Exp Hematol 25:187–192

Vargas-Roig LM, Fanelli MA, Lopez LA, Gago FE, Tello O, Aznar JC, Ciocca DR (1997) Heat shock proteins and cell proliferation in human breast cancer biopsy samples. Cancer Detect Prev 21:441–451

Vaziri H, Benchimol S (1998) Reconstitution of telomerase activity in normal human cells leads to elongation of telomeres and extended replicative life span. Curr Biol 8:279–282

Venable ME, Lee JY, Smyth MJ, Bielawska A, Obeid LM (1995) Role of ceramide in cellular senescence. J Biol Chem 270:30701–30708

Vogt M, Haggblom C, Yeargin J, Christiansen-Weber T, Haas M (1998) Independent induction of senescence by p16INK4a and p21CIP1 in spontaneously immortalized human fibroblasts. Cell Growth Differ 9:139–146

Wadhwa R, Kaul SC, Ikawa Y, Sugimoto Y (1991) Protein markers for cellular mortality and immortality. Mutat Res 256:243–254

Wadhwa R, Kaul SC, Sugimoto Y, Mitsui Y (1993a) Induction of cellular senescence by transfection of cytosolic mortalin cDNA in NIH 3T3 cells. J Biol Chem 268:22239–22242

Wadhwa R, Kaul SC, Mitsui Y, Sugomoto Y (1993b) Differential subcellular distribution of mortalin in mortal and immortal mouse and human fibroblasts. Exp Cell Res 207:442–448

Wadhwa R, Kaul SC, Sugimoto Y, Mitsui Y (1993c) Spontaneous immortalization of mouse fibroblasts involves structural changes in senescence inducing protein, mortalin. Biochem Biophys Res Commun 197:202–206

Wadhwa R, Kaul SC, Ikawa Y, Sugimoto Y (1993d) Identification of a novel member of mouse hsp70 family. Its association with cellular mortal phenotype. J Biol Chem 268:6615–6621

Wadhwa R, Pereira-Smith OM, Reddel RR, Sugimoto Y, Mitsui Y, Kaul SC (1995) Correlation between complementation group for immortality and the cellular distribution of mortalin. Exp Cell Res 216:101–106

Wadhwa R, Akiyama S, Sugihara T, Redel RR, Mitsui Y, Kaul SC (1996) Genetic differences between the pancytosolic and perinuclear forms of murine mortalin. Exp Cell Res 226:381–386

Wareham KA, Lyon MF, Glenister PH, Williams ED (1987) Age related reactivation of an X-linked gene. Nature 327:725–727

Warner HR (1997) Aging and regulation of apoptosis. Curr Top Cell Regul 35:107–121

Watanabe M, Suzuki K, Kodama S, Suguhara T (1995) Normal human cells at confluence get heat resistance by efficient accumulation of hsp72 in nucleus. Carcinogenesis 16:2373–2380

Watanabe Y, Lee SW, Detmar M, Ajioka I, Dvorak HF (1997) Vascular permeability factor/vascular endothelial growth factor (VPF/VEGF) delays and induces escape from senescence in human dermal microvascular endothelial cells. Oncogene 14:2025–2032

Webster TJ, Naylor DJ, Hartman DJ, Hoj PB, Hoogenraad NJ (1994) cDNA cloning and efficient mitochondrial import of pre-mtHSP70 from rat liver. DNA Cell Biol 13:1213–1220

Wei YQ, Zhao X, Kariya Y, Fukata H, Teshigawara K, Uchida A (1994) Induction of apoptosis by quercitin: involvement of heat shock protein. Cancer Res 54:4952–4957

Witkowsky JA (1980) Dr. Carrel's immortal cells. Med Hist 24:129–142

Wong H, Riabowol K (1996) Differential CDK-inhibitor gene expression in aging human diploid fibroblasts. Exp Gerontol 31:311–325

Xiao H, Hasegqwa T, Miyaishi O, Ohkusu K, Isobe KI (1997) Sodium butyrate induces NIH3T3 cells to senescence-like state and enhances promoter activity of p21WAF/CIPI in p53-independent manner. Biochem Biophys Res Commun 237:457–460

Xu HJ, Zhou Y, Ji W, Perng GS,, Kruzelock R, Kong CT, Bast RC, Mills GB, Li J, Hu SX (1997) Re-expression of the retinoblastoma protein in tumor cells induces senecence and telomerase inhibition. Oncogene 20:2589–2596

Yano M, Naito Z, Tanaka S, Asano G (1996) Expression and roles of heat shock proteins in human breast cancer. Jpn J Cancer Res 87:908–915

Yufu Y, Nishimura J, Nawata H (1992) High constitutive expression of heat shock protein 90 alpha in human acute leukemia cells. Leuk Res 16:597–605 Zglinicki T von, Saretzki G (1997) Molecular mechanisms of senescence in cell culture. Z Gerontol Geriatr 30:24–28

Zhu W, Roma P, Pirillo A, Pellegatta F, Catapano AL (1996) Human endothelial cells exposed to oxidized LDL express hsp70 only when proliferating. Arterioscler Thromb Vasc Biol 16:1104–1111

Cell Immortality: Maintenance of Cell Division Potential

C. Bernstein[1], H. Bernstein[1] and C. Payne[1]

1
Introduction

Cell immortality refers to the ability to reproduce indefinitely. This property does not imply constancy of genetic information from generation to generation, since mutation coupled with natural selection and genetic drift may cause genetic changes over successive generations. Furthermore, cells of a germ line ordinarily undergo periodic recombination with cells of other germ lines causing additional genetic change. Nevertheless, all extant cells reflect the ability to reproduce indefinitely, since the ancestry of each cell presumably traces back, in an unbroken lineage for over 3 billion years, to the origin of life. As pointed out by Avise (1993), it is not actually cells which are immortal, but cell lineages.

Cell immortality depends on the maintenance of cell division potential. This potential may be lost due to cellular damage, terminal differentiation as in a nerve cell, programmed cell death (e.g. selective cell loss through apoptosis during development) or as the result of aging and death of the organism. The maintenance of cell division potential thus depends on the avoidance and repair of cellular damage and, in obligate sexual organisms, transmission through the germ line rather than the somatic line.

Cell immortality occurs in two natural contexts:
(1) vegetative reproduction in bacteria, protozoans, fungi, plants and some invertebrates, and
(2) the germ line of sexually reproducing organisms. In addition,
(3) certain types of mammalian somatic cells (e.g. virally transformed cells and cancer cells), although not naturally immortal since they do not outlast the multicellular organism of which they are a part, nevertheless are regarded as potentially immortal since they can often maintain their division capacity when serially cultivated in vitro. The ability to avoid and repair cellular damage is important for maintaining cell division potential in all three of these contexts.

---

[1] Department of Microbiology and Immunology, College of Medicine, University of Arizona, Tucson, Arizona 85724, USA.

Progress in Molecular and Subcellular Biology, Vol. 24
A. Macieira-Coelho
© Springer-Verlag Berlin Heidelberg 1999

2
Avoiding Damage by Coping with Cellular Stresses

Cells have evolved sophisticated mechanisms for avoiding the damage that arises from specific stresses that they ordinarily encounter. Prominent among these is the stress caused by oxidative metabolism.

2.1
Oxidative Stress Response

The maintenance of life in an aerobic environment appears to require resolution of a basic conflict between the need to maintain genetic information and the benefit of using oxygen to produce energy (which has destructive side effects). An apparently unavoidable byproduct of normal respiratory metabolism is the generation of reactive oxygen species (ROS) produced from molecular oxygen. ROS include the superoxide radical (O_2^-), hydrogen peroxide (H_2O_2) and the hydroxyl radical (·OH) (Henle and Linn 1997). ROS cause DNA, protein and membrane damage. The most damaging of the ROS is the hydroxyl radical produced from H_2O_2 by the Fenton reaction. ROS are also formed by catabolic oxidases such as xanthine oxidase, anabolic processes such as nucleoside reduction, and defense processes such as phagocytosis. Respiring cells have adapted to this intrinsic problem by evolving several systems for protecting against this damage.

2.1.1
Oxidative Stress Response in Bacteria

Respiring bacteria express several enzymes that protect against oxidative damage. *Escherichia coli* has two types of superoxide dismutase (SOD), manganese-containing SOD (MnSOD) encoded by the *sodA* gene and iron-containing SOD (FeSOD) encoded by *sodB*, and two forms of catalase, one encoded by *katG* and the other by *katE* and *katF*. SOD catalyzes conversion of the superoxide radical into hydrogen peroxide and oxygen, and catalase converts the hydrogen peroxide into molecular oxygen and water.

Low levels of H_2O_2 exposure to *E. coli* and other bacteria (see review by Demple and Amabile-Cuevas, 1991), elicits an adaptive response that counteracts the toxicity of much higher H_2O_2 levels. A separate adaptive response mitigates cell killing by superoxide generating agents. In *E. coli*, these two adaptive responses correspond to two „stimulons"–one for H_2O_2 and one for O_2^-. Each stimulon has about 40 genes, but for most of these genes neither their function nor the manner of their regulation is understood. However, within the H_2O_2 stimulon the *oxyR* gene product mediates the H_2O_2-inducible expression of eight genes. These constitute the *oxyR* regulon. Within the O_2^- stimulon, the *soxR* gene is thought to induce expression of the *soxS* gene, and the *soxS* gene induces expression of nine additional genes. These constitute the *soxRS* regulon. The *soxRS* regulon responds to O_2^- and nitric oxide, but not H_2O_2. The *oxyR* and *soxRS* gene products exert positive transcriptional control of their respective regulon genes, and the activation of these responses greatly increases cellular resistance to oxidative agents. Two of the genes in the *oxyR* regulon are the genes for the peroxide destroying enzymes catalase *(katG)* and NADPH-dependent alkylhydroper-oxidase *(ahpFC)*. The *soxRS* regulon provides a multilevel defense by increasing the synthesis of MnSOD *(sodA)*, the DNA repair enzyme endonu-clease IV *(nfo)*, and glucose-6-phosphate dehydrogenase *(zwf)*, an enzyme that produces reducing equivalents in the cell in the form of NADPH.

2.1.2
Oxidative Stress Response in Mammalian Cells

In mammalian cells, oxidative stress responses are regulated by the transcription factors nuclear transcription factor κB (NF-κB) and activator protein 1 (AP-1) (Schulze-Osthoff et al. 1997). NF-κB plays a major role in the activation of many genes, especially defensive genes during immune and inflammatory responses. NF-κB is involved in the regulation (or prevention) of programmed cell death (apoptosis) (Baeuerle and Baltimore 1996). Several recent studies have shown that inhibition of NF-κB potentiates apoptosis (see, for example, Payne et al. 1998). Ap-1 is a complex of the Jun and Fos transcription factor families. It is tightly regulated by redox processes at both the transcriptional and post transcriptional levels (Schulze-Osthoff et al. 1997). Another set of proteins, the heat shock proteins, also play an important role in the cell's response to oxidative stress (Yaagoubi et al. 1998). Other proteins involved in protection against oxidative stress are catalase, SOD, glutathione, glucose-6-phosphate dehydrogenase, thioredoxin and Bcl-2 (Pandolfi et al. 1995; Payne et al. 1995a).

Ataxia telangiectasia (AT) is a recessive genetic disorder. Affected individuals have cerebellar ataxia, immunodeficiency, specific developmental defects, disposition to cancer and acute radiosensitivity. AT is due to mutation in the ATM gene. The ATM gene product acts as a sensor of reactive oxygen species and/or oxidative damage to DNA (see review by Rotman and Shiloh, 1997). This protein is thought to induce signalling through multiple pathways, thereby coordinating stress responses with cell cycle checkpoint control and repair of oxidative damage. Loss of ATM function limits the repair of oxidative damage that occurs under normal physiological conditions, ultimately leading to apoptosis of particularly sensitive cells, such as neurons and thymocytes. Gene products thought to be under ATM control include NF-κB, PARP (see Sect. 3.1.1) and p53 (see Sects. 7.1 and 7.2).

Despite the available protective processes, the average amount of oxidative DNA damage occurring per cell per day is estimated to be about 10 000 in humans, and in rat, with a higher metabolic rate, about 100 000 (Ames et al. 1993). Most of this damage is single-strand, mainly altered bases such as

thymine glycol and 8-hydroxyguanosine. About 2 % of the damage is probably double-strand, such as interstrand cross-links and double-strand breaks
(Massie et al. 1972), which are more difficult to repair than single-strand
damage. These observations imply that oxidative DNA damage is a serious
threat to cell survival. Oxidation is not the only source of damage to DNA.
Lindahl (1993) reviewed evidence that spontaneous hydrolysis and nonenzy-
matic methylation of DNA occur at significant rates in vivo. He estimated
that in each human cell 2 000 to 10 000 DNA purine bases turn over every
day due to hydrolytic depurination and subsequent repair.

2.2
Most Genes Help Maintain Cell Division Potential

The number of genes present in the genome of bacteria is typically a few
thousand (e.g. *E. coli* has about 4288 genes according to Blattner et al. 1997)
and in humans it is estimated to be about 80 000 (Hartl and Jones 1998).
Presumably, most of the genes in any organism contain information that
provides an adaptive benefit under some circumstance that the organism
ordinarily encounters. Many gene products are typically concerned with
acquiring and using external resources (food) for energy and as components of cellular structures, with replicating the genetic material, and with
defense against disease and predation. These genes, like the oxidative stress
response genes discussed above, are important, in a broad sense, for maintaining cell division potential. However, the genetic systems for responding
to oxidative stress are emphasized here because they are among the most
well studied responses to a specific, but ubiquitous, challenge to cell survival.

3
Repair of Damage

Proteins, lipids, carbohydrates and nucleic acids are all damaged by reactive oxygen species and other endogenous reactive agents (Pacifici and
Davies 1991). Oxidatively damaged proteins and lipids are subject to
degradation and some repair reactions. Macromolecules such as RNA and
protein, and structures such as internal and external membranes, ribosomes and the cytoskeleton are also subject to damage. In principle, damaged molecules (other than DNA) and macromolecular structures (e.g.
organelles) can be replaced as long as the genetic information of the cell is
intact, material and energy resources are available, and the damage sustained does not overwhelm the capacity of the cell to cope. Proteins ordinarily have a short half-life (e.g. about 3 days in mouse liver, Barrows and
Kokkonen 1987). There has been no compelling evidence that protein
damage is of general importance in cellular aging (Gafni 1990). In contrast,
even one or a few unrepaired damages in DNA can have drastic deleterious

effects on cell function, including cell death, by blocking chromosome replication, by interfering with the transcription of vital genes, or by causing replication errors leading to deleterious mutation. Kunz et al. (1998) estimated that, in yeast, at least 60 % of spontaneous single base pair substitution and deletion mutations may be caused by synthesis across DNA damage in the template strand.

DNA is usually not degraded unless the amount of damage is overwhelming. Rather it is subject to several complex repair processes. Damage to DNA appears to be a serious threat to cell immortality because of DNA's unique role as the bearer of the cell's genetic information and the requirement for it to faithfully replicate at each cell division. Thus DNA repair processes are central to maintaining cell immortality.

3.1
Repair of DNA Damage

Several important DNA repair processes require informational redundancy to replace damaged information. For single-strand damage the source of the redundant information can be the intact complementary strand of DNA. For double-strand damage, that is, damage which affects both strands of DNA at or near the same position, the complementary strand of DNA cannot serve as a template for obtaining correct information. Thus, repair of single-strand damage is relatively easy. These can be handled by excision repair, where the damage is excised and the correct information to fill in the resulting gap is obtained from the intact partner strand, which serves as a template for repair synthesis. There are two types of excision repair: base excision repair and nucleotide excision repair. These are discussed below. There are also two modes of repair of double strand damages, one that involves recombinational repair and the other a process of end joining. Recombinational repair involves accurate transfer of information to the damaged DNA from an intact homolog. An inaccurate process of repair involves end joining to restore double-strand breaks. These two processes are also discussed below. Another important form of repair, DNA mismatch repair, is discussed briefly in Section 7.

3.1.1
Base Excision Repair

Base excision repair is initiated by the action of a DNA glycosylase which catalyzes the hydrolysis of the N-glycosylic bond linking bases to the deoxyribose backbone (Lehman 1998). Different DNA glycosylases are specific for different types of altered bases. Examples of specific glycosylases found in mammalian cells are uracil-DNA glycosylase, 8-oxoguanine glycosylase, hypoxanthine glycosylase, thymine-glycol glycosylase and 3-methyladenine glycosylase. The DNA glycosylase reaction leaves either an apyrimidinic or apurinic site in the DNA, referred to as AP sites. AP sites can also be formed by spontaneous (i.e. heat-induced at body temperature) hydrolysis of the glycosylic bond. Once an AP site is formed, its removal requires the action of one or more nucleases. These enzymes carry out the incision or nicking of DNA at AP sites by catalyzing the hydrolysis of phosphodiester bonds. After the initial incision, an exonuclease can remove the AP site, as well as additional neighboring nucleotides. The resulting single-strand gap may then be filled in by a DNA polymerase which uses the undamaged strand as a template. The final phosphodiester bond can be sealed by DNA ligase.

In most eukaryote organisms (but not in yeast or prokaryotes), one of the immediate reactions to the occurrence of oxidative and other types of damage is addition of poly(ADP-ribosyl) groups to selected proteins by poly(ADP-ribose) polymerase (PARP). The proteins which are modified include histones, other chromosomal proteins, a Ca^{2+}-Mg^{2+}-activated endonuclease, and PARP itself. This reaction is strongly stimulated by DNA single- or double-strand breaks. PARP appears to be part of a protein complex that promotes DNA base excision repair (Masson et al. 1998). The source of the ADP-ribosyl groups is NAD^+. Burkle et al. (1995) have shown that PARP activity correlates with average (and maximum) life span in comparisons of different mammalian species. This suggests that base excision repair is a factor in promoting increased life span of the organism, and, by implication, in maintaining cell immortality.

3.1.2
Nucleotide Excision Repair

A wide variety of different types of damage in DNA which result in large distortions of the double helical structure are removed by nucleotide excision repair. This process requires the protein products of about 30 genes. The mechanism involves the recognition of the damage, the removal of the single-stranded segment of the DNA containing the damage and then the replacement of the damaged segment by synthesis employing the undamaged strand as template (Sancar 1994; Lehman 1998). In humans, deficiencies in nucleotide excision repair give rise to genetic disorders; the most well studied of these is xeroderma pigmentosum (XP). The XP phenotype can arise from mutations in each of seven genes, XPA to XPG; there is also a separate class known as XP variants.

XP is an autosomal recessive disease. Individuals who are homozygous for an XP gene are hypersensitive to UV-irradiation and have a high incidence of skin cancer and a moderately increased incidence of internal cancers. XP cells are defective in the repair of DNA containing pyrimidine dimers and a variety of other bulky damages. Satoh et al. (1993) presented evidence that accumulation of endogenous oxidative damage in cellular DNA from XP patients contributes to the increased frequency of internal cancers and the neural degeneration occurring in serious cases of the syndrome. Another inherited disease in humans, Cockayne syndrome (CS) is also caused by a defect in the nucleotide excision repair of DNA damage in transcriptionally active DNA (Hanawalt 1994). This condition has several features of premature aging.

3.1.3
Recombinational Repair

Recombinational repair is a form of repair which permits damage in one double-stranded DNA molecule to be replaced by an undamaged DNA sequence from another double-stranded DNA molecule. Recombinational repair has been shown to occur in a wide range of prokaryotes and eukaryotes. Bernstein et al. (1987) and Bernstein and Bernstein (1991) have reviewed the experiments demonstrating that recombinational repair is prevalent among viruses, bacteria, and fungi and that it occurs in *Drosophila*, mammals and plants. Further, recombinational repair is effective against DNA damage introduced by a wide variety of agents and, in particular, is very efficient in overcoming double-strand damage in both *E. coli* and the yeast *S. cerevisiae*. When *S. cerevisiae* cells were recombination proficient and diploid, it took 35 double-strand breaks or 120 cross-links to induce an average of one lethal hit per cell (Resnick and Martin 1976; Magana-Schwencke et al. 1982). However, when recombination genes were defective, an average of about two double-strand breaks or just one cross-link per cell was sufficient to introduce a lethal hit.

3.1.4
DNA Cross-link Repair

In *E. coli,* both nucleotide excision repair and recombinational repair are necessary to remove DNA interstrand cross-links (Sladek et al. 1989). The process appears to involve dual incisions by excision nuclease in one strand on both sides of the cross-linked base. This is followed by transfer of an undamaged strand from a homologous DNA (mediated by RecA protein, see below, Sects. 5.2–5.3) to fill in the gap generated by the dual incisions. This intermediate is then subjected to a second round of dual incisions by the excision nuclease in the second strand to eliminate the cross-link and to generate a gapped duplex. This remaining gap can be filled in by using the previously repaired strand as template.

In *Drosophila*, Harris et al. (1996) isolated a novel protein, the product of the *mus308* gene, which is essential for cross-link repair. This protein has domains homologous to DNA helicase and DNA polymerase I. These properties suggest that there may be energetic coupling of repair synthesis with branch migration across the region containing the cross-link, necessitated by the barrier to branch migration represented by the cross-link.

Bessho et al. (1997) showed that cross-link repair in humans is initiated by action of nucleotide excision nuclease. They found that in contrast to monoadducts, which are removed by dual incisions bracketing the damage, the cross-link causes dual incisions, both 5' to the cross-link in one of the two strands. The net result is formation of a 22- to 28-nucleotide-long gap immediately 5' to the cross-link. This gap may act as a recombinogenic signal to promote subsequent cross-link removal.

3.1.5
Double-strand Break Repair in Somatic Cells of Mammals

An important pathway of repair of double-strand breaks in mammals is mediated by DNA-dependent protein kinase (DNA-PK) (Chu 1997). DNA-PK is composed of the Ku protein heterodimer, which binds to DNA ends, together with a catalytic subunit DNA-PKcs whose kinase activity is activated when it is recruited by DNA-bound Ku. The alignment of the broken ends is mediated by base pairing in regions of microhomology involving 1–6 base pairs between the otherwise non-homologous strands. This alignment then leads to subsequent steps in the end-joining reaction. Ordinarily, the original sequence in the region of the break is not preserved and a variety of mutations are introduced such as deletions and insertions. This form of inaccurate repair appears to be the predominant mechanism for handling double-strand breaks in mammalian somatic cells. However, mammalian cells also appear to be able to use a recombinational repair process for repairing double-strand breaks (Park 1995).

4
Survival Strategy of Vegetative Cell Populations

Vegetatively growing populations of bacteria can be regarded as potentially immortal as long as nutrient resources are available. However, even in such populations there may be constant attrition due to DNA damage and deleterious mutation. Cutler (1972) carried out experiments in which cells of *Escherichia coli* were attached to a nitrocellulose filter. They were then allowed to grow and form daughter cells. Those daughter cells, which were unattached, could be eluted into the media. The rate of duplication of the cells originally attached declined steadily until at six days no further duplication occurred. This decline could be explained by accumulation of DNA damage and/or mutation. Similarly, yeast mother cells have an average lifespan of about 25 daughter cell buds (Austriaco and Guarente 1997).

Flowering plants generally reproduce sexually with gametes formed by a meiotic process and gamete fusion as the prelude to embryogenesis and seed formation. Plants, unlike most animals, are also able to regenerate complete new individuals of similar genetic constitution from vegetative parts. Meristematic buds or excised pieces of tissue can be induced to propagate when placed in the appropriate environment. Apparently, plant vegetative cell lines can be maintained indefinitely (Osborne 1985). These lines probably maintain themselves by a strategy of replacement, where cells with lethal unrepaired DNA damage or expressed deleterious mutations are replaced by replication of non-defective cells. Nevertheless, in some plant tissues DNA damage may accumulate. Cheah and Osborne (1978) showed that, in dry seeds, fragmentation of nuclear DNA occurs with time.

Most forest trees live for at least 100 years, many of them for more than 300 years, and a few for more than 1 000 years. Clonal tree species may occupy a site for several thousand years. Most of the tree is dead, and only a thin shell of dividing cells (cambium) is present around the trunk and in the leaves. Cutler (1976) has noted that a tree actually represents a free-living clone of cells in which selective removal of cells with irreversible accumulated damage is constantly occurring. He observed that one would not expect to find old cells in a tree any more than one would find old cells in a growing culture of bacteria. The evidence reviewed in this section suggests that some proliferating cell populations can cope with unrepaired DNA damage by a replacement strategy which can be maintained indefinitely as long as nutrient resources are available and the level of unrepaired DNA damage is not excessive.

5
Potential Immortality of Germ Cells

The germ line of a eukaryote is the cellular lineage that connects successive meioses. Since the germ line is a key feature of sexual reproduction, it is useful for further discussion to define sexual reproduction so as to clarify its key components. Such a definition, to be general, should encompass sexual processes in both eukaryotes and prokarotes. *Sexual reproduction is the process by which genetic material (usually DNA) from two separate parents is brought together in a common cytoplasm where recombination of the genetic material ordinarily occurs, followed by the passage of the recombined genome(s) to progeny.* Thus sexual reproduction has two key elements:

(1) recombination, in the sense of the breakage and exchange of DNA between two homologous chromosomes, and
(2) outcrossing, in the sense that homologous chromosomes from two different individuals come together in the same cell (Bernstein et al. 1985).

In obligate sexual organisms, the germ line is potentially immortal in contrast to the somatic line. This feature of the germ line presumably is due to adaptations present in the germline that are absent in the somatic line. Meiosis is a process unique to the germ line, and it appears to be an adaptation specifically for promoting recombination. The term „recombination" in this context refers to general recombination, as distinct from site specific recombination (as occurs during prophage integration, replicative transposition, or immunoglobulin gene segment rearrangement) which operates by a different mechanism.

Some diploid organisms form gametes by meiosis, but then undergo self-fertilization or automixis (a process in which two haploid products of meiosis fuse to form a diploid zygote). These processes are common in plants and invertebrates. Self-fertilization and automixis are not strictly sexual processes since they lack the outcrossing aspect of sex. However, some of the organisms which undergo these asexual processes are also facultatively sexual. In the discussion which follows we assume that meiosis serves essentially the same function in all organisms in which it occurs, and that the germ line in facultative or obligate self-fertilizing and automictic eukaryotes is the cell lineage that connects successive meioses.

5.1
Repair of DNA Damage in Germ Cells

Although the germ line is characterized by periodic events of meiosis, during the intervals beween such events, cell divisions are ordinarily by mitosis. Cells of the germ line are presumably capable of the same types of DNA repair processes that occur in vegetative cells, discussed in section 3, above. For instance, Corominas and Mesquita (1989) have demonstrated a relatively high level of PARP activity in premeiotic and meiotic spermatocytes, an indication of base excision repair activity. (They also suggest that the germinal cells may have effective mechanisms of replenishment of NAD used by PARP as a substrate, since these cells are remarkably resistant to suicidal NAD depletion.) Fraga et al. (1990) measured the accumulation of one type of oxidatively damaged DNA base, 8-hydroxy-2'- deoxyguanosine, in various cells of the rat. While the average accumulation in the rat kidney was 80 residues per cell per day, there was no detectable accumulation in the testes. This non-accumulation in the testes can be interpreted as a reflection of efficient base excision repair. In addition to the repair mechanisms also available to somatic cells, gamete formation occurs by meiosis, a process in which non-sister homologous chromosomes pair and recombine. Evidence is next reviewed indicating that meiotic recombination, and hence meiosis, is an adaptation for DNA repair. To put this evidence in proper context, however, it is first necessary to discuss the relationship of recombination processes in prokaryotes and eukaryotes.

5.2
Recombination in Eukaryotes Probably Evolved from Recombination in Prokaryotes

Several workers have reviewed evidence suggesting that the processes of recombination in prokaryotes and eukaryotes share a common ancestry. Dougherty (1955) was possibly the first to conclude that „the evolution of sexuality as it exists today was the result of a single phylogenetic sequence," based on the fact that recombination in bacteriophage and in bacteria, on the one hand, and meiotic recombination in eukaryotes, on the other hand, seemed to share fundamental similarities. Stahl (1979), reviewing material in the succeeding 24 years, concluded that despite numerous differences in detail the „similarities in recombination in creatures as diverse as the phage and fungi are impressive".

In the succeeding years, there has been considerable work on the biochemistry of recombination with a focus on the RecA protein of *E. coli*. This protein catalyzes the key steps in recombination of ATP-dependent homologous DNA pairing and strand exchange. Homologs of the *E. coli recA* gene have been found in over 60 bacterial species and in bacteriophage T4. This suggests that RecA-catalyzed recombination is very common in the prokaryotic world. Since about 1992 there has been much work on RecA homologs in eukaryotes indicating that RecA homologs play a key role in meiotic recombination in fungi and vertebrates. Recently, a RecA homolog was found in humans that showed 30 % amino acid sequence identity with the *E. coli* RecA protein. RecA homologs in yeast and humans form helical filaments with DNA, very similar to those formed by *E. coli* RecA. The conservation of the DNA-RecA filament structure led Benson et al. (1994) to conclude that the Rec A protein has been conserved from bacteria to man. These findings suggest that eukaryotic meiotic recombination and recombination processes in extant bacteria are probably both derived from a recombination process present in a common ancestor that existed before the divergence of prokaryotes and eukaryotes, at least 1.8 billion years ago (Doolittle et al. 1989), and presumably have the same adaptive function.

5.3
Homologs of the *E. coli* RecA Protein Have a Key Role in Meiotic Recombination

Substantial evidence indicates that RecA homologs have a central role in meiotic recombination. In the yeast *Saccharomyces cerevisiae*, the *rad51* and *dmc1* genes are homologs of the *recA* gene of *E. coli*. The Rad51 and Dmc1 proteins appear to share redundant functions, since recombination is reduced only a few fold in *dmc1* and *rad51* single mutants, but *dmc1* and *rad51* double mutants are profoundly defective in meiotic recombination (Camerini-Otero and Hsieh 1995). The Rad51 protein acts during both mitosis and meiosis, whereas the Dmc1 protein acts specifically during meiosis. Both Rad51 and Dmc1 proteins are employed in repair of double-strand breaks (Camerini-Otero and Hsieh 1995). Sung (1994) showed that the Rad51 protein, like *E. coli* recA protein, catalyzes ATP-dependent homologous DNA pairing and strand exchange. Ogawa et al. (1993) showed that the yeast Rad51 protein polymerizes on double-stranded DNA to form a helical filament nearly identical to that formed by the *E. coli* RecA protein. Story et al. (1993) showed that the Dmc1 protein has an overall similarity to *E. coli* RecA protein in tertiary structure. These findings suggest that the yeast Rad51 and the bacterial RecA proteins are similar in function as well as in sequence.

In the mouse, Morita et al. (1993) showed that a gene homologous to the *E. coli recA* gene is expressed at a high level in the testes. Shinohara et al. (1993) also showed that the mouse *recA* homolog is expressed at a high level in testis and ovary, and suggested that the gene is involved in meiotic recombination. The expression of a *recA* gene homolog has also been demonstrated in chicken testis and ovary and in human testes. All these proteins have about a 30 % amino acid identity over their central core region with *E. coli* RecA (reviewed in Camerini-Otero and Hsieh 1995). The human RecA homolog carries out the hallmark reactions of *E. coli* RecA protein, including DNA-dependent hydrolysis of ATP, renaturation of complementary strands, homologous pairing of a single strand with duplex DNA, and strand exchange (Gupta et al. 1997). A pair of *recA* gene homologs of the lily, *LIM15* and *RAD51*, which are, in turn, homologs of the yeast *dmc1* and *rad51* genes, were found by Terasawa et al. (1995). The *LIM15* gene in lily is specifically expressed in meiotic prophase during microsporogenesis.

The evidence reviewed so far indicates that eukaryotic RecA protein homologs active at meiosis and *E. coli* RecA protein function in a similar manner. Cox (1993) showed that DNA repair is probably the principal adaptive function of the RecA protein and its homologs. He pointed out that RecA protein binding is largely limited to regions in the DNA containing suitable nucleation sites, especially single-strand gaps. A variety of DNA damage also facilitates nucleation. Although RecA protein does not specifically recognize such damage as binding sites, the structural perturbations caused provide favorable nucleation sites.

Cox also discussed the very high use of ATP in RecA catalyzed reactions. About 100 ATPs are hydrolyzed for every base pair of heteroduplex DNA generated by RecA mediated strand exchange. He reviewed evidence that the energy released by ATP hydrolysis is used specifically to allow the strand exchange process to traverse damaged regions of DNA. He argued that this use of ATP is readily understood as an adaptation for repair of DNA damage. Thus the evidence reviewed by Cox indicates that the adaptive function of RecA homologs acting during meiosis is DNA repair.

5.4
The Adaptive Function of Meiosis in *Paramecium*

A second line of evidence suggesting that the adaptive function of meiotic recombination is DNA repair comes from studies with the ciliate protozoan *Paramecium tetraurelia*. *P. tetraurelia* has a diploid micronucleus and a polyploid macronucleus, the macronucleus containing about 800 to 1500 copies of the genome. The macronuclear DNA expresses cellular functions while the micronucleus contains the germline DNA. This organism may reproduce asexually by binary fission, or by a meiotic process: either conjugation, which is a kind of outcrossing sex, or automixis, which is a kind of self-fertilization. In the asexual phase, during which cell divisions are by mitosis rather than meiosis, there is a gradual loss of vitality, or clonal aging.

Transplantation experiments by Aufderheide (1987) illuminate the cause of clonal aging. If the macronuclei of clonally young paramecia are injected into paramecia of a standard type, the life span of the recipient is prolonged. In contrast, cytoplasmic transfer from young paramecia does not prolong the life span of the recipient. These experiments suggest that the macronucleus, rather than the cytoplasm, determines clonal aging. Studies by Smith-Sonneborn (1979), Holmes and Holmes (1986) and Gilley and Blackburn (1994) indicate that during clonal aging there is a dramatic increase in DNA damage. When clonally aged paramecia are allowed to undergo meiosis in association with either automixis or conjugation, the progeny are rejuvenated. Furthermore, the old macronucleus disintegrates and a new macronucleus is reconstituted by replication of the micronuclear DNA, the DNA which had recently undergone meiosis. The new macronucleus has little, if any, DNA damage. These findings suggest that clonal aging is due to a progressive increase in DNA damage, and that rejuvenation is due to repair of DNA damages during meiosis. A line of clonally aging paramecia will die out after about 200 fissions if it does not undergo automixis or conjugation. Therefore, in this system, meiosis appears to be an adaptation for removing DNA damages in the germ line to allow survival.

5.5
Specialized Strategies for Avoiding Damage to Germ Cells

As described above, the germ line is characterized by periodic meiotic events which allow an especially effective repair process (recombinational repair between non-sister homologs) to occur just prior to formation of the germ cells. Here we discuss other characteristics of the germ line that are apparently further adaptations for avoiding damage to the DNA of the germ cells. Human egg cells, with a diameter of about 100µm, are quite large compared with typical somatic cells which are, on average, about 10 µm in diameter. Human egg cells, thus, have about a thousand-fold greater mass than a somatic cell. An insect egg is even larger than a human one, having a mass that is about one-million-fold greater than a human somatic cell. The much greater mass of egg cells than somatic cells implies that the metabolic activity used to provision the egg is greater than that ordinarily used by the average somatic cell. Considering the importance of the egg for production of progeny and the large amount of metabolic activity needed to provision the egg, it raises the question of how the egg avoids excess damage to its DNA from the oxidative byproducts of this metabolism.

The answer appears to be that much of the material within an egg depends on the synthetic activity of other cells (Alberts et al. 1994). For instance, the egg cells of some insects such as *Drosophila* are surrounded by nurse cells to which they are connected by cytoplasmic bridges. These nurse cells provide much of the material of the egg cell including much of the protein, mRNA and ribosomes. Presumably these nurse cells are subject to oxidative damage that would otherwise befall the egg cell. Each nurse cell contains hundreds of genome copies, and this redundancy of genetic information may allow the nurse cells to carry out their provisioning even in the face of considerable oxidative DNA damage.

Vertebrate oocytes are surrounded by follicle cells which occur in a layer around the oocyte and are connected to it by gap junctions (Alberts et al. 1994). These junctions are not large enough to allow transfer of bulky macromolecules, but they can transmit precursor molecules. For chickens, amphibians and insects, yolk proteins are accumulated by the egg. These proteins are made by liver (or equivalent) cells that secrete the yolk proteins into the blood from which they are taken up by the oocytes. Thus it appears that oocytes of many organisms are protected from oxidative damage, while storing up provisions to nurture the potential zygote in its initial embryonic growth, by relying on other cells to provide resources.

The evidence reviewed in this Section and in Section 5.4 above suggests that the formation of germ cells is adapted both for avoiding and repairing damage to the nuclear DNA.

5.6
Vegetative Strategy for Mitochondria (and Chloroplasts) in the Germ Line

As pointed out by Avise (1993), a molecular-level analogue of cellular replacement ('molecular replacement') facilitates the purging of both DNA damage and deleterious mutations in cytoplasmic genomes (mitochondria and chloroplasts) of germ line cells. While nuclear genes exist as single allelic copies per gamete, there are thousands of mitochondrial DNA (mtDNA) molecules in most cells, and several hundred thousand may co-habit a mature oocyte. The many mtDNAs in oocytes appear to stem from a vastly smaller pool of mtDNA molecules that must survive a process of replicative segregation in earlier cytokinetic divisions of the germ line lineage, since most heterogeneity of mtDNA is distributed among, rather than within individuals, and this implies there are relative mtDNA population bottlenecks in germ lines. That is, the mtDNA, which is generally solely transmitted through the oocyte, is of only one or a few genotypes. Avise concluded that the mtDNA molecules that survive and replicate to populate a mature oocyte appear to have been rather scrupulously screened by natural selection for replicative capacity and functional competence in the germ-cell lineages they inhabit. This is equivalent, at the molecular level, to a vegetative cell replacement strategy for cell lineages that do not have a sexual cycle. While the mitochondria and chloroplasts within cells follow a vegetative replacement strategy, they, like vegetatively replicating bacteria and yeast, utilize homologous DNA recombinational repair to repair DNA damage due to oxidation or environmental stresses (Cerutti et al. 1995).

5.7
Dolly the Cloned Lamb, and Cumulina the Cloned Mouse

The evidence reviewed above indicates that germ cells that have undergone meiosis are relatively free of DNA damage, and thus should be able to give rise to viable offspring with high probability. At lower probability, somatic cells may also be sufficiently free of DNA damage to be able to produce viable progeny.

Wilmut et al. (1997) derived eight viable lambs, one (named Dolly) from a donor cell of a mature sheep, four from donor embryo-derived cells, and three from donor fetal fibroblast cells. For Dolly, the lamb derived from a somatic cell of a mature sheep, they first created 277 fused couplets (enucleated oocytes fused to donor cells), using donor cells in their 3rd to 6th passage, cultured from the mammary gland of a six year old pregnant ewe. A morula or blastocyst formed from only 11.7 % of the cultured fused couplets from the mature ewe. A morula/blastocyst was formed 27 % to 39 % of the time from couplets derived from embryo or fetal fibroblast cells. Some of the morula/fibroblasts, implanted in recipient ewes, were able to form fetuses detectable by ultrasound at 50–60 days. Subsequently, 62 % of these fetuses were lost, a significantly greater proportion than the estimate of 6 % after natural (meiosis based) mating in sheep.

Similar results were obtained with mice. Wakayama et al. (1998) derived 10 healthy mice, the first of which was named Cumulina, from donor nuclei of differentiated, non-replicating granulosa cells of mature mice injected into enucleated recipient mouse oocytes. Cumulina and the other nine cloned healthy newborn mice were produced from 800 injected oocytes which had produced embryos and were transplanted into foster mother recipient mice. In a number of cloning experiments by these authors, the rates of successful implantation of embryos was 57–71 %, formation of fetuses was 5–16 %, and full term development was 2–3 %.

The low survival of fetuses in both sheep and mice could have three possible explanations. First, injuries introduced by the experimental manipulation of the embryos before implantation might harm further development. Wilmut et al. (1997) refer to work showing that even unreconstructed embryos experience increased prenatal loss after manipulation or culture. Second, the differentiated donor nuclei, transplanted into the recipient oocytes, have to be reprogrammed to set the developmental clock to zero and errors in this process may have deleterious effects on fetal development. Third, the low survival of fetuses may reflect the greater amount of DNA damage in donor somatic cells (than in meiotically produced gametes), which can give rise to deleterious mutations when replicating after transfer to an oocyte. The relative importance of these three factors is not known.

6
DNA Damage, Cellular Mortality and Aging

The impact of oxidative damage on aging of the organism was assessed by creating two transgenic *Drosophila* which carried three, rather than the usual two, copies either of the gene encoding SOD or the gene encoding catalase. In both of these *Drosophila* there was little or no effect on life span. However, when both enzymes were allowed to overexpress in transgenic *Drosophila*, as much as a 30 % increase in mean and maximum life spans was observed (Orr and Sohal 1994). Despite greater physical activity and oxygen consumption, these flies suffered less oxidative damage to protein and DNA (Sohal et al. 1995). In another study, Dudas and Arking (1995) showed that in a long-lived strain of *Drosophila* the delayed onset of senescence was associated with an upregulation of the antioxidant genes encoding Cu/Zn superoxide dismutase, catalase and xanthine dehydrogenase. These observations suggest that cellular oxidative damage promotes senescence and limits life span, whereas the avoidance of such damage delays senescence and increases life span. Parkes et al. (1998) created two transgenic lines of *Drosophila* which overexpressed human SOD1 in their motor neurons. These flies have a life span that is 40 % longer than their normal counterparts. It is not clear why this effect occurs, whereas there is no effect on adult *Drosophila* life span when SOD levels have been increased broadly throughout many tissues (e.g. Sohal et al. 1995). Nevertheless, these results suggest that the nervous system is a major target for ROS induced damage which limits life spans.

In the microscopic worm *Caenorhabditis elegans*, the *age-1* mutation increases maximum lifespan by 110 %. These worms have elevated levels of Cu/Zn superoxide dismutase and catalase. *Age-1* mutant worms also have increased thermotolerance, resistance to heat shock and resistance to UV irradiation (Lithgow and Kirkwood 1996). These results suggest that resistance to oxidative stress, as well as to other stresses, extends the life span.

Hart and Setlow (1974) tested the ability of skin fibroblasts of shrew, mouse, rat, hamster, cow, elephant and human to perform unscheduled DNA synthesis after UV irradiation (as a measure of DNA repair, and probably mainly nucleotide excision repair). They found that the extent of unscheduled DNA synthesis increased systematically with life span. Since this original study, at least 15 further studies, reviewed by Bernstein and Bernstein (1991), have been carried out on mammals to examine this correlation. In all but two of these studies, DNA repair levels correlated with lifespan. These observations suggest that excision repair removes DNA damage that would otherwise contribute to aging.

In mammals, restriction of caloric intake retards age-associated changes, and extends maximum life span. These benefits are associated with lowered steady-state levels of oxidative stress and damage (Sohal and Weindruch 1996).

Several human genetic syndromes with features of accelerated aging appear to have increased DNA damage or defective DNA repair. Examples of such conditions are ataxia telangiectasia (see Sect. 2.1.2) and Cockayne syndrome (see Sect. 3.1.2).

6.1
Coping with Unrepaired DNA Damage in Somatic Cell Populations

Despite the processes for avoiding DNA damage and repairing that which occurs, a fraction of DNA damage appears to remain unrepaired. Different organisms or even different tissues within the same organism appear to use different strategies for coping with unrepaired DNA damage. At one extreme, the strategy of continuous replication and replacement of damaged cells may allow a cell population to replicate more rapidly than the unrepaired damages can deplete the population (see Sect. 4 where this strategy is described for bacteria, yeast and higher plants). This survival strategy also seems to be used by some somatic cell populations in mammals. Where it is used in multicellular organisms, the replication and replacement strategy is vulnerable to mutationally variant clonal populations that may proliferate abnormally. Human cancer (see Sect. 7, below) is an example of such a problem.

Mammals and insects have substantial cell populations that are non-dividing. Apparently, these cell populations rely primarily on DNA repair for coping with DNA damage. This strategy avoids the high costs of cell turnover associated with the replication and replacement strategy, as well as the problem of mutations that occur during replication, which can cause abnormal proliferation. However, there is a cost in terms of accumulation of unrepaired damage and the consequent aging of the cells in the population. Humans apparently use a mixture of strategies in different somatic cell populations for coping with DNA damage. For instance, brain and muscle are composed in large part of non-dividing cells, whereas hematopoietic cells in the bone marrow replicate rapidly.

6.1.1
Non-Dividing Somatic Cell Populations

Organs such as brain, muscle and liver are subject to some of the more conspicuous progressive declines in function that characterize human aging. Bernstein and Bernstein (1991) reviewed evidence in the brain that endogenous DNA damage accumulates with age, mRNA synthesis declines, and general protein synthesis as well as synthesis of critical specific proteins is reduced. Similar evidence was also reviewed for muscle, showing that DNA damage accumulates with age, general protein synthesis as well as synthesis of some specific proteins declines, cellular structure deteriorates, and cells die. Also, for liver, evidence indicates that cell loss accompanied by degenerative changes in morphology and function occur with age.

Holmes et al. (1992) reviewed studies on the accumulation of DNA damage in mammalian muscle, brain and liver that were reported in the period 1971–1991. In summary, 4 studies on muscle, 9 studies on brain, and 14 on liver reported accumulation of DNA damage with age. In most of these studies, the type of damage measured was single-strand breaks. More recently, Mandavilli and Rao (1996) found an increase in the number of DNA single- and double-strand breaks in neuronal cells of the rat cerebral cortex with age. They estimated that neurons in young 4-day-old rats had about 3 000 single-strand breaks which increased to about 7400 in neurons of old rats greater than 2 years of age. Double-strand breaks increased from about 156 to 600 in young versus old rats. They suggested that gradual accumulation of DNA damage with age could be a primary reason for the breakdown of metabolic machinery and could lead to the eventual senescence and death of the neuron. Cai et al. (1996) reported that indigenous DNA adducts increase in rat brain with age. Some of these were identified as malondialdehyde adducts of dGMP. They suggested that this accumulation may contribute to cerebral aging. Higami et al. (1994) found a significant increase in single-strand breaks/alkali-labile sites in old compared with young rat liver hepatocytes, using a new technique which measures DNA damage in individual cells. These numerous studies indicate that long-lived, non-dividing, differentiated cells accumulate DNA damage with time. This damage may account for many of the progressive declines in function that define aging and prevent cell immortality of many mammalian somatic cell lineages.

6.1.2
Dividing Somatic Cell Populations

As indicated above in Section 4, in a population of rapidly dividing cells, damaged cells can be replaced by the duplication of other undamaged cells in the population. In particular, if the rate of occurrence of unrepaired lethal DNA damage is low compared with the cell generation time (that is, less than 0.5 unrepaired lethal DNA damages per cell per generation), then populations of such cells may not experience the progressive decline in function that defines aging. Thus, within an organism such a population of cells, as a whole, may maintain its cell division potential for the life of the organism, even though individual cells might die.

According to the extensive data summarized by Buetow (1985), bone marrow and hematopoietic cells do not decrease in numbers with age in the guinea pig and mouse. Mori et al. (1986) studied hematopoietic stem cells in young and old humans ranging in age from 28–95 years. They found no difference in the concentration of granulocyte-macrophage progenitor cells, although they did observe a significant decrease in the concentration of erythrocyte progenitor cells in elderly people. The turnover time for replacing the hematopoietic bone marrow cells of the mouse is only 1–2 days (Bowman 1985). As reviewed by Harrison (1979), in the mouse, erythropoietic stem cells have a very large capacity for self-renewal. When erythrocyte production by marrow stem cells from young and old individuals were compared, no significant differences were found. These findings suggest that little, if any, erythropoietic stem cell proliferative capacity is exhausted in the mouse by a lifetime of normal functioning. Thus, at least for this type of rapidly replicating cell, loss of cell division potential may not be a significant determinant of aging. However, some types of dividing cell (e.g. human fibroblasts) lose division potential with age as indicated by their reduced life span when serially cultured in vitro (see Sect. 7.1). Senescent fibroblasts and keratinocytes appear to be present in skin from old donors and accumulate with age in vivo (Dimri et al. 1995). These changes may contribute to organismal aging and may be reflected, for example, in reduced capacity for wound healing with age.

Rapidly dividing cells should be vulnerable to accumulation of mutations which arise by errors of replication. Akiyama et al. (1995) have presented evidence that hematopoietic stem cells accumulate mutations as people age from birth to 96 years. The accumulation of mutations in replicating somatic cells is widely regarded as the cause of cancer (see Sect. 7).

7
Potential Immortality of Cancer Cells

As pointed out by Ames et al. (1993), a critical factor in mutagenesis is cell division. When a cell divides, unrepaired DNA damage can give rise to a mutation by replication past the damage in the template strand. Thus, a major factor in mutagenesis, and hence carcinogenesis, is the cell division rate in the precursors of tumor cells.

Jackson and Loeb (1998) have pointed out that in order for a cell to acquire the requisite number of mutations (estimated to be six to ten) to become malignant, the cell probably would have to mutate at an early stage of carcinogenesis to a mutator phenotype. Such a phenotype is known to arise from mutations in genes encoding the enzymes catalyzing DNA base mismatch repair. Mutations in the DNA repair enzyme, DNA polymerase β, also give rise to a mutator phenotype associated with cancer (Wang et al. 1992). Destabilization of simple sequence repeats, called microsatellites, is a marker of the mutator phenotype, and as reviewed by Jackson and Loeb (1998), many different types of cancer are characterized by such instability.

Normal human fibroblasts are able to undergo only a finite number of cell divisions even under optimum culture conditions, before becoming senescent, i.e. non-proliferative (Hayflick 1965). Another constraint on indefinite cell division is programmed cell death (apoptosis). However, many human tumors contain variant cells that possess indefinite proliferative potential in vitro (Edington et al. 1995) due to mutations in protooncogenes and tumor suppressor genes which remove signals for senescence and apoptosis.

7.1
Escape from Cellular Senescence in Relation to Cancer

The molecular cause of cellular senescence is not known. Numerous genomic modifications occur as cells proliferate in vitro (reviewed by Macieira-Coelho 1993), and the last cell divisions before entering the non-dividing state are characterized by a chaotic behavior in the distribution of DNA between daughter cells (Macieira-Coelho 1995). It has been suggested that an increase of DNA errors (damage and mutations) (Weirich-Schwaiger et al. 1994) or telomere erosion (Sect. 8) could be the basis of cellular senescence. Cultured untransformed human fibroblasts from young and old healthy individuals and from individuals with premature senescence syndromes (Werner syndrome, Cockayne syndrome, ataxia telangiectasia and Down syndrome) were examined by Weirich-Schwaiger et al. (1994). The cultured cells from healthy old individuals and from individuals with the premature aging syndromes all had a shorter life span in culture than cells of young healthy individuals. Furthermore, there was a striking parallelism between reduced maximal life span of the cultured cells and increased cell cycle duration, on the one hand, and elevated levels of chromosomal breaks, higher incidence of micronuclei (a measure of genetic instability), and diminished DNA repair capacity, on the other hand. These results suggested a causal relationship between DNA errors and senescence of cultured cells.

After a finite number of cell divisions in culture, cellular senescence progresses to a point where one or more signal pathways inhibit cell cycling. Two possible protein inhibitors whose levels rise with proliferative life span are p16 (Hara et al. 1996) and $p21^{sdi1/WAF1}$ (Noda et al. 1994). These proteins inhibit the cyclin-dependent kinases CDK4/6 and CDK2, which are required for passage through and exit from the G1 phase of the cell cycle (Sherr and Roberts 1995). Another protein, p24, which is related to $p21^{sdi1/WAF1}$ (Mazars and Jat 1997), may be a third inhibitor. A consequence of the action of these inhibitors is the failure of cells to phosphorylate the product of the retinoblastoma sensitivity gene *pRb*, a tumor suppressor gene, (Stein et al. 1990). In its active unphosphorylated form pRb sequesters transcription factors needed for G_1-S progression (Weinberg 1995). As might be expected, escape from senescence is frequently associated with deregulation of the Rb pathway, either directly due to loss of Rb itself, or indirectly as a result of loss of p16. As reviewed by Jansen-Durr (1998) oncoprotein E7 and E6 encoded by the human papilloma virus may allow escape from cellular senescence by inactivating a further CDK inhibitor $p27^{KIP1}$ or inactivating p53, respectively.

Like pRb, p53 is also the product of a tumor suppressor gene. Escape from senescence is associated with loss of p53 function (Wynford-Thomas 1996). The transcriptional transactivation function of p53 is activated as human fibroblasts senesce (Bond et al. 1996), possibly in response to telemere shortening. Activation of p53 is a strong inducer of the CDK inhibitor $p21^{sdi1/WAF1}$ (El-Deiry et al. 1993). Thus, p53 is a possible link between the cell division clock and cell cycle inhibition. In support of this idea, Gire and Wynford-Thomas (1998) have shown that when anti-p53 antibodies are microinjected into senescent human fibroblasts both DNA synthesis and cell division are reinitiated. Thus the maintenance of fibroblast senescence appears to be critically dependent on functional p53.

7.2
Escape from Apoptosis in Relation to Cancer

Mammalian cells are able to undergo a process of programmed cell death (apoptosis) in response to a variety of stimuli. Apoptosis may have evolved in multicellular organisms as a natural defense against the retention of cells with unrepaired DNA damage, to benefit the overall survival of the organism through a form of cellular euthanasia (altruistic cell suicide) of damaged cells. The benefit to the organism arises from the fact that replication past unrepaired DNA damage in the template will lead, with high probability, to mutations, accumulation of which may lead to cancer. Apoptosis can limit carcinogenesis and cancer progression (Cotter et al 1990).

Bedi et al. (1995) investigated whether colon cancer involves an altered susceptibility to apoptosis by examining colorectal epithelium from normal mucosa, flat mucosa and adenomas from familial adenomatous polyposis patients, as well as sporadic adenomas and carcinomas. They found that the transformation of colorectal epithelium to carcinomas was associated with a progressive inhibition of apoptosis. Also Arai and Kino (1995) found a significantly lower frequency of spontaneous apoptosis in villous adenomas, known to develop into large tumors, when compared with tubular adenomas.

As reviewed by Cheah (1990), bile salts have been implicated in the etiology of colon cancer by numerous studies. Sodium deoxycholate (NaDOC), the bile salt present in highest concentration in the human colon and feces (Govers et al. 1996), induces apoptosis in human colonic epithelial cells (Payne et al. 1995b) at a concentration comparable to those accompanying a high-fat diet (Stadler et al. 1988). Payne et al. (1995b) and Garewal et al. (1996) found that within the normal appearing mucosa of patients with a history of colon cancer, cells are relatively resistant to bile acid induced apoptosis. On the basis of these findings they hypothesized that excessive apoptosis induced by bile salts, produced in response to a high fat diet, leads to the selection of apoptosis-resistant cells. If cells are apoptosis-resistant, they may not die when they experience unrepaired DNA damage. This could lead, upon replication of the damaged DNA, to mutation, leading to colon cancer and a potentially immortalized cell lineage.

Bax is a member of the Bcl2 family of apoptosis genes transactivated by p53 in response to DNA damage (White 1996). It is thought that p53 can induce growth arrest as described above (Sect. 7.1), or apoptosis, depending apparently on the cell type and degree of damage. Both activities are potentially involved in tumor suppression (White 1996). Colon cancers that exhibit microsatellite instability are often mutated in a particular repetitive sequence in the *Bax* gene (Rampino et al. 1997). Since the only known function of *Bax* is to promote apoptosis (White 1996), inactivation of *Bax* by mutation may result in diminished capacity to trigger apoptosis in response to DNA damage, despite the presence of functional p53 protein. As a result, cells with extensive DNA damage may persist and have an increased probability of generating mutations during subsequent DNA replication cycles, and result in potentially immortalized cell lineages.

8
Other Factors Affecting Immortality of Cell Lines

Human tissues which are normally part of an immortal cell line (germ line testis cells) or a non-senescing population (intestine) have detectable telomerase levels, as do most tumors examined, while tissues that senesce, including heart, brain, placenta, liver, skeletal muscle and prostate, have no detectable level of telomerase (Meyerson et al. 1997). Human cell lines, transfected with plasmids that express telomerase and cause increased telomere lengths, become immortalized (Bodnar et al. 1998). Mild oxidative stress can cause shortening of telomeres and inhibit proliferation of fibroblasts in a manner which is very similar to that seen in senescent cells (Zglinicki et al. 1995). The mechanism by which telomerase or longer telomeres confers immortality in human cell lines is unknown.

Telomerase itself is apparently not essential for immortality or for normal telomere length. As reviewed by Lansdorp (1997), mice without any telomerase function (double telomerase-knockout mice) were able to give rise to six generations of mice with normal telomere length in the germline, and had cells which were efficiently transformed into immortal cell lines or into in vivo tumor-forming cell lines. He also pointed out that approximately one quarter of immortalized human cell lines lack detectable telomerase activity. Larger telomeres may not be essential for immortality of cell lines. In the yeast *Saccharomyces cerevisiae,* longer telomere lengths were correlated with shorter life spans of mother cells (Austriaco and Guarente 1997).

Other factors have been reported which cause immortalization of cell lines in culture. One of these involves immortalized hepatocyte cell lines, cloned from mice expressing the transgene *AT-cyto-MET,* which encodes the cytoplasmic portion of Met, where Met is the hepatocyte growth factor receptor and is a transmembrane tyrosine kinase (Amicone et al. 1997).

Although some of these immortalized hepatocyte cell lines do not express cyto-Met after 25 or 30 generations, it was concluded that the initial expression of cyto-Met was required to allow immortalization, since initial hepatocyte cell lines lacking cyto-Met could not form immortalized clones. Two short DNA sequences, isolated from the cytoplasm of mouse tumor cells, induced immortalization of human lymphocytes after transfection in vitro (Abken et al. 1993). The mechanism by which these factors cause immortalization of cell lines is unknown.

9
Conclusions

The two major natural contexts in which potential cell immortality is maintained are vegetative reproduction and the germ line of eukaryotes. A ubiquitous challenge to cell survival is formation of reactive oxygen species as a byproduct of normal oxidative metabolism which produces DNA damage. During vegetative reproduction, cell division potential appears to be maintained in large part by the avoidance and repair of DNA damage, as well as by a selection process in which damaged cells are lost from the dividing population. The germ line of eukaryotes is characterized by meiosis, the process by which germ cells are formed. Meiosis appears to be an adaptation for promoting recombinational repair of DNA damage, especially double-strand damage. Aging in humans and other organisms appears to involve accumulation of DNA damage and a consequent reduction in function in non-dividing or slowly dividing cells, such as those of the brain, muscle and liver. Functional declines with age also appear to occur in dividing cells, and such cells are vulnerable to accumulation of mutations leading to reduced function and cancer. The potential immortality of some cancer cell lines arises when constraints to cell division and survival, which normally apply to somatic cells of multicellular eukaryotes, are relaxed.

Acknowledgments. This work was supported by NIH Program Project Grant #CA72008, and Arizona Disease Control Research Commission grant #9621.

References

Abken H, Hegger R, Butzler C, Willecke K (1993) Short DNA sequences from the cytoplasm of mouse tumor cells induce immortalization of human lymphocytes in vitro. Proc Natl Acad Sci USA 90:6518–6522

Akiyama M, Kyoizumi S, Hirai Y, Kusunoki Y, Iwamoto KS, Nakamura N (1995) Mutation frequency in human blood cells increases with age. Mutat Res 338:141–149

Alberts B, Bray D, Lewis J, Raff M, Roberts K, Watson JD (1994) Molecular biology of the cell. Garland, London

Ames BN, Shigenaga MK, Hagen TM (1993) Oxidants, antioxidants, and the degenerative diseases of aging. Proc Natl Acad Sci USA 90:7915–7922

Amicone L, Spagnoli FM, Spath G, Giordano S, Tommasini C, Bernadini S, De Luca V, Della Rocca C, Weiss MC, Comoglio PM, Tripodi M (1997) Transgenic expression in the liver of truncated Met blocks apoptosis and permits immortalization of hepatocytes. EMBO J 16:495–503

Arai T, Kino I (1995) Role of apoptosis in modulation of the growth of human colorectal tubular and villous adenomas. J Pathol 176:37–44

Aufderheide KJ (1987) Clonal aging in *Paramecium tetraurelia*. II. Evidence of functional changes in the macronucleus with age. Mech Ageing Dev 37:265–279

Austriaco NR, Guarente LP (1997) Changes of telomere length cause reciprocal changes in the life span of mother cells in *Saccharomyces cerevisiae*. Proc Natl Acad Sci USA 94:9768–9772

Avise JC (1993) The evolutionary biology of aging, sexual reproduction, and DNA repair. Evolution 47:1293–1301

Baeuerle PA, Baltimore D (1996) NF-κB: Ten years after. Cell 87:13–20

Barrows CH, Kokkonen G (1987) The effect of age and diet on the cellular protein synthesis of liver of male mice. Age 10:54–57

Bedi A, Pasricha PJ, Akhtar AJ, Barber JP, Bedi GC, Giardiello FM, Zehnbauer BA, Hamilton SR, Jones RJ (1995) Inhibition of apoptosis during development of colorectal cancer. Cancer Res 55:1811–1816

Benson FE, Stasiak A, West SC (1994) Purification and characterization of the human Rad51 protein, an analogue of *E. coli* RecA. EMBO J 13:5764–5771

Bernstein C, Bernstein H (1991) Aging, Sex and DNA Repair. Academic Press, New York

Bernstein H, Byerly HC, Hopf FA, Michod RE (1985) Genetic damage, mutation and the evolution of sex. Science 229:1277–1281

Bernstein H, Hopf FA, Michod RE (1987) The molecular basis of the evolution of sex. Adv Genet 24:323–370

Bessho T, Mu D, Sancar A (1997) Initiation of DNA interstrand cross-link repair in humans: the nucleotide excision repair system makes dual incisions 5' to the cross-linked base and removes a 22- to 28-nucleotide-long damage-free strand. Mol Cell Biol 17:6822–6830

Blattner FR, Plunkett G, Bloch CA, Perna NT, Burland V, Riley M, Collado-Vides J, Glasner JD, Rode CK, Mayhew GF, Gregor J, Davis NW, Kirkpatrick HA, Goeden MA, Rose DJ Mau B, Shao Y (1997) The complete genome sequence of *Escherichia coli* K-12. Science 277:1453–1474

Bodnar AG, Ouellette M, Frolkis M, Holt SE, Chiu C-P, Morin GB, Harley CB, Shay JW, Lichtsteiner S, Wright WE (1998) Extension of life span by introduction of telomerase into normal human cells. Science 279:349–352

Bond J, Haughton M, Blaydes J, Gire V, Wynford-Thomas D, Wyllie F (1996) Evidence that transcriptional activation by p53 plays a direct role in the induction of cellular senescence. Oncogene 13:2097–2104

Bowman PD (1985) Aging and the cell cycle in vivo and in vitro. In: Cristofolo VJ, Adelman RC, Roth GS (eds) Handbook of cell biology of aging. CRC Press, Boca Raton, Florida, pp 117–136

Buetow DE (1985) Cell numbers vs. age in mammalian tissues and organs. In: Cristofolo VJ, Adelman RC, Roth GS (eds) Handbook of cell biology of aging. CRC Press, Boca Raton, Florida, pp 1–115

Burkle A, Grube K, Muller M, Wolf I, Heller B, Kupper JH (1995) Poly(ADP-ribose) polymerase: correlation of enzyme activity with the life span of mammalian species and use of a dominant negative version to elucidate biological functions of poly(ADP ribosyl)ation. In: Cutler RG, Packer L, Bertram J, Mori A (eds) Oxidative stress and aging. Birkhauser, Basle/Switzerland, pp 111–121

Cai Q, Tian L, Wei H (1996) Age-dependent increase of indigenous DNA adducts in rat brain is associated with a lipid peroxidation product. Exp Gerontol 31:387–392

Camerini-Otero RD, Hsieh P (1995) Homologous recombination proteins in prokaryotes and eukaryotes. Annu Rev Genet 29:509–552

Cerutti H, Johnson AM, Boynton JE, Gillham NW (1995) Inhibition of chloroplast DNA recombination and repair by dominant negative mutants of *Escherichia coli* RecA. Molec Cell Biol 15:3003–3011

Cheah KSE, Osborne DJ (1978) DNA lesions occur with loss of viability in embryos of aging rye seed. Nature (Lond) 272:593–599

Cheah PY (1990) Hypotheses for the etiology of colorectal cancer. An overview. Nutr and Cancer 14:5–13

Chu G (1997) Double-strand break repair. J Biol Chem 272:24097–24100

Corominas M, Mesquita C (1989) DNA damage does not induce lethal depletion of NAD during chicken spermatogenesis. In: Jacobson MK, Jacobson EL (eds) ADP-ribose transfer reactions: Mechanisms and biological significance. Springer Berlin Heidelberg New York, pp 326–329

Cotter TG, Lennon SV, Glynn JG, Martin SJ (1990) Cell death via apoptosis and its relationship to growth, development and differentiation of both tumour and normal cells. Anticancer Res 10:1153–1160

Cox MM (1993) Relating biochemistry to biology: how the recombinational repair function of RecA protein is manifested in its molecular properties. Bioessays 15:617–623

Cutler RG (1972) Transcription of reiterated DNA sequence classes throughout the life span of the mouse. Adv Gerontol Res 4:219–321

Cutler RG (1976) Nature of aging and life maintenance processes. Interdiscip Topics Gerontol 9:83–133

Demple B, Amabile-Cuevas CF (1991) Redox redux: the control of oxidative stress responses. Cell 67:837–839

Dimri GP, Lee X, Basile G, Acosta M, Scott G, Roskelley C, Medrano EE, Linskens M, Rubelj I, Pereira-Smith O, Peacocke M, Campisi J (1995) A biomarker that identifies senescent human cells in culture and in aging skin in vivo. Proc Natl Acad Sci USA 92:9363–9367

Doolittle RF, Anderson KL, Feng DF (1989) Estimating the prokaryote-eukaryote divergence time from protein sequences. In: Fernholm B, Bremer K, Jornvall H (eds) The hierarchy of life, Elsevier, Amsterdam, pp 73–85

Dougherty EC (1955) Comparative evolution and the origin of sexuality. Syst Zool 4:145–190

Dudas SP, Arking R (1995) A coordinate upregulation of antioxidant gene activities is associated with the delayed onset of senescence in a long-lived strain of *Drosophila*. J Gerontol Biol Sci 50A: B117-B127

Edington KG, Loughran OP, Berry IJ, Parkinson EK (1995) Cellular immortality: a late event in the progression of human squamous cell carcinoma of the head and neck associated with p53 alteration and a high frequency of allele loss. Mol Carcinog 13:254–265

El-Deiry WS, Tokino T, Velculescu VE, Levy DB, Parsons R, Trent JM, Lin D, Mercer WE, Kinsler KW, Vogelstein B (1993) WAF1, a potential mediator of p53 tumour suppression. Cell 75:817–825

Fraga CG, Shigenaga MK, Park JW, Degan P, Ames BN (1990) Oxidative damage to DNA during aging: 8-hydroxy-2'-deoxyguanosine in rat organ DNA and urine. Proc Natl Acad Sci USA 87:4533–4537

Gafni A (1990) Age-related effects in enzyme metabolism and catalysis. Rev Biol Res Aging 4:315–336

Garewal H, Bernstein H, Bernstein C, Sampliner R, Payne C (1996) Reduced bile-acid induced apoptosis in „normal" colorectal mucosa: a potential biomarker for cancer risk. Cancer Res 56:1480–1483

Gilley D, Blackburn EH (1994) Lack of telomere shortening during senescence in *Paramecium*. Proc Natl Acad Sci USA 91:1955–1958

Gire V, Wynford-Thomas D (1998) Reinitiation of DNA synthesis and cell division in senescent human fibroblasts by microinjection of anti-p53 antibodies. Mol Cell Biol 18:1611–1621

Govers MJAP, Termont DSML, Lapre JA, Kleibeuker JH, Vonk RJ, Van der Meer R (1996) Calcium in milk products precipitates intestinal fatty acids and secondary bile acids and thus inhibits colonic cytotoxicity in humans. Cancer Res 56:3270–3275

Gupta RC, Bazemore LR, Golub EI, Radding CM (1997) Activities of human recombination protein Rad51. Proc Natl Acad Sci USA 94:463–468

Hanawalt C (1994) Transcription-coupled repair and human disease. Science 266: 1957–1958

Hara E, Smith R, Parry D, Tahara H, Stone S, Peters G (1996) Regulation of p16^{CDKN2} expression and its implications for cell immortalization and senescence. Mol Cell Biol 16:859–867

Harper JL, White J (1974) The demography of plants. Annu Rev Ecol Syst 5:419–463

Harris PV, Mazina OM, Leonhardt EA, Case RB, Boyd JB, Burtis KC (1996) Molecular cloning of *Drosophila mus308*, a gene involved in DNA cross-link repair with homology to prokaryotic DNA polymerase I genes. Mol Cell Biol 16:5764–5771

Harrison DE (1979) Proliferative capacity of erythropoietic stem cell lines and aging: an overview. Mech Ageing Dev 9:409–426

Hart RW, Setlow RB (1974) Correlation between deoxyribonucleic acid excision-repair and life span in a number of mammalian species. Proc Natl Acad Sci USA 71:2169–2173

Hartl DL, Jones EW (1998) Genetics: principles and analysis. Sudbury, MA, USA Hayflick L (1965) The limited in vitro lifetime of human diploid cell strains. Exp Cell Res 37:614–636

Henle ES, Linn S (1997) Formation, prevention, and repair of DNA damage by iron/hydrogen peroxide. J Biol Chem 272:19095–19098

Higami Y, Shimokawa I, OkimotoT, Ikeda T (1994) An age-related increase in the basal level of DNA damage and DNA vulnerability to oxygen radicals in the individual hepatocytes of male F344 rats. Mutat Res 316:59–67

Holmes GE, Holmes NR (1986) Accumulation of DNA damages in aging *Paramecium tetraurelia*. Mol Gen Genet 204:108–114

Holmes GE, Bernstein C, Bernstein H (1992) Oxidative and other DNA damages as the basis of aging: a review. Mutat Res 275:305–315

Jackson AL, Loeb LA (1998) The mutation rate and cancer. Genetics 148:1483–1490

Jansen-Durr P (1998) The making and the breaking of senescence: changes of gene expression during cellular aging and immortilalization. Exp Gerontol 33:291–301

Kunz BA, Ramachandran K, Vonarx EJ (1998) DNA sequence analysis of spontaneous mutagenesis in *Saccharomyces cerevisiae*. Genetics 148:1491–1505

Lansdorp PM (1997) Lessons from mice without telomerase. J Cell Biol 139:309–312

Lehmann AR (1998) Dual functions of DNA repair genes: molecular, cellular, and clinical implications. BioEssays 20:146–155

Lindahl T (1993) Instability and decay of the primary structure of DNA. Nature 362:709–715

Lithgow GJ, Kirkwood TBL (1996) Mechanisms and evolution of aging. Science 273: 80

Macieira-Coelho A (1993) Contributions made by the studies of cells in vitro for understanding of the mechanisms of aging. Exp Gerontol 28:1–16

Macieira-Coelho A (1995) Chaos in DNA partition during the last mitoses of the proliferative life span of human fibroblasts. FEBS Lett 358:126–128

Magana-Schwencke N, Henriques JAP, Chanet R, Moustacchi E (1982) The fate of 8-methoxypsoralen photo-induced crosslinks in nuclear and mitochondrial yeast DNA: Comparison of wild-type and repair deficient strains. Proc Natl Acad Sci USA 79:1722–1726

Mandavilli BS, Rao KS (1996) Accumulation of DNA damage in aging neurons occurs through a mechanism other than apoptosis. J Neurochem 67:1559–1565

Massie HR, Samis HV, Baird MB (1972) The kinetics of degradation of DNA and RNA by H_2O_2. Biochim Biophys Acta 272:539–548

Masson M, Niedergang C, Schreiber V, Muller S, Menissier-deMurcia J, DeMurcia G (1998) XRCC1 is specifically associated with poly(ADP-ribose) polymerase and negatively regulates its activity following DNA damage. Mol Cell Biol 18:3563–3571

Mazars GR, Jat PS (1997) Expression of p24, a novel p21$^{Waf1/cip1/Sdi1}$ related protein, correlates with measurement of the finite proliferative potential of rodent embryo fibroblasts. Proc Natl Acad Sci USA 94:151–156

Meyerson M, Counter CM, Eaton EN, Ellisen LW, Steiner P, Caddle SD, Ziaugra L, Beijersbergen RL, Davidoff MJ, Liu Q, Bacchetti S, Haber DA, Weinberg RA (1997) *hEST2*, the putative human telomerase catalytic subunit gene, is up-regulated in tumor cells during immortalization. Cell 90:785–795

Mori M, Tanaka A, Sato N (1986) Hematopoietic stem cells in elderly people. Mech Ageing Dev 37:41–47

Morita T, Yoshimura Y, Yamamoto A, Murata K, Mori M, Yamamoto H, Matsushiro A (1993) A mouse homolog of the *Escherichia coli recA* and *Saccharomyces cerevisiae RAD51* genes. Proc Natl Acad Sci USA 90:6577–6580

Noda A, Ning Y, Venable SF, Pereira-Smith OM, Smith JR (1994) Cloning of senescent cell-derived inhibitors of DNA synthesis using an expression screen. Exp Cell Res 211:90–98

Ogawa T, Yu X, Shinohara A, Egelman EH (1993) Similarity of the yeast RAD51 filament to the bacterial RecA filament. Science 259:1896–1899

Orr WC, Sohal RS (1994) Extension of life span by overexpression of superoxide dismutase and catalase in *Drosophila melanogaster*. Science 263:1128–1130

Osborne DJ (1985) Annual plants. Interdiscip Top Gerontol 21:247–262

Pacifici RE, Davies KJA (1991) Protein, lipid and DNA repair systems in oxidative stress: the free radical theory of aging revisited. Gerontology 37:166–180

Pandolfi PP, Sonati F, Rivi R, Mason P, Grosveld F, Luzzatto L (1995) Targeted disruption of the housekeeping gene encoding glucose 6-phosphate dehydrogenase (G6PD): G6PD is dispendable for pentose synthesis but essential for defense against oxidative stress. EMBO J 14:5209–5215

Park MS (1995) Expression of human RAD52 confers resistance to ionizing radiation in mammalian cells. J Biol Chem 270:15467–15470

Parkes TL, Elia AJ, Dickenson D, Hilliker AJ, Phillips JB, Boulianne GL (1998) Extension of *Drosophila* lifespan by overexpression of human DOD1 in motor neurons. Nature Genetics 19:171–174

Payne CM, Bernstein C, Bernstein H (1995a) Apoptosis overview emphasizing the role of oxidative stress, DNA damage and signal transduction pathways. Leukemia and Lymphoma 19:43–93

Payne CM, Bernstein H, Bernstein C, Garewal H (1995b) Role of apoptosis in biology and pathology: Resistance to apoptosis in colon carcinogenesis. Ultrastruct Pathol 19:221–248

Payne CM, Crowley C, Washo-Stultz D, Briehl M, Bernstein H, Bernstein C, Beard S, Holubec H, Warneke J (1998) The stress-response proteins poly(ADP-ribose) polymerase and NF-κB protect against bile salt-induced apoptosis. Cell Death Differ 5:623–636

Rampino N, Yamamoto H, Ionov Y, Li Y, Sawai H., Reed JC, Perucho M (1997) Somatic frameshift mutations in the BAX gene in colon cancers of the microsatellite mutator phenotype. Science 275:967–969

Resnick MA, Martin P (1976) The repair of double-strand breaks in the nuclear DNA of *Saccharomyces cerevisiae* and its genetic control. Mol Gen Genet 143:119–129

Rotman G, Shiloh Y (1997) Ataxia-telangiectasia: is ATM a sensor of oxidative damage and stress? BioEssays 19:911–917

Sancar A (1994) Mechanisms of DNA excision repair. Science 266:1954–1956

Satoh MS, Jones CJ, Wood RD, Lindahl T (1993) DNA excision-repair defect of xeroderma pigmentosum prevents removal of a class of oxygen free radical-induced base lesions. Proc Natl Acad Sci USA 90:6335–6339

Schulze-Osthoff K, Bauer MKA, Vogt M, Wesselborg S (1997) Oxidative stress and signal transduction. Int J Vitam Nutr Res 67:336–342

Sherr CJ, Roberts JM (1995) Inhibitors of mammalian G_1 cyclin-dependent kinases. Genes Dev 9:1149–1163

Shinohara A, Ogawa H, MatsudaY, Ushio N, Ikeo K, Ogawa T (1993) Cloning of human, mouseand fission yeast recombination genes homologous to RAD51 and recA. Nat Genet 4:239–243

Sladek FM, Munn MM, Rupp WD, Howard-Flanders P (1989) In vitro repair of psoralen-DNA cross-links by RecA, UvrABC, and the 5'-exonuclease of DNA polymerase I. J Biol Chem 264:6755–6765

Smith-Sonneborn J (1979) DNA repair and longevity assurance in *Paramecium tetraurelia*. Science 203:1115–1117

Sohal RS, Weindurch R (1996) Oxidative stress, caloric restriction, and aging. Science 273:59–63

Sohal RS, Agarwal A, Agarwal S, Orr WC (1995) Simultaneous overexpression of copper and zinc-containing superoxide dismutase and catalase retards age-related oxidative damage and increases metabolic potential in *Drosophila melanogaster*. J Biol Chem 270:15671- 15674

Stadler J, Stern HS, Yeung KS, McGuire V, Furrer R, Marcon N, Bruce WR (1988) Effect of high fat consumption on cell proliferation activity of colorectal mucosa and on soluble fecal bile acids. Gut 29:1326–1331

Stahl FW (1979) Genetic recombination: thinking about it in phage and fungi. Freeman, San Francisco

Stein GH, Beeson M, Gordon L (1990) Failure to phosphorylate the retinoblastoma gene product in senescent human fibroblasts. Science 249:666–669

Story RM, Bishop DK, Kleckner N, Steitz TA (1993) Structural relationship of bacterial RecA proteins to recombination proteins from bacteriophage T4 and yeast. Science 259:1892–1896

Sung P (1994) Catalysis of ATP-dependent homologous DNA pairing and strand exchange by yeast RAD51 protein. Science 265:1241–1243

Terasawa M, Shinohara A, Hotta Y, Ogawa H, Ogawa T (1995) Localization of RecA-like recombination protein on chromosomes of the lily at various meiotic stages. Genes Dev 9:925–934

Wakayama T, Perry ACF, Zucotti M, Johnson KR, Yanagimachi R (1998) Full-term development of mice from enucleated oocytes injected with cumulus cell nuclei. Nature 394:369–374

Wang L, Patel U, Ghosh L, Banerjee S (1992) DNA polymerase β mutations in human colorectal cancer. Cancer Res 52:4824–4827

Weinberg RA (1995) The retinoblastoma protein and cell cycle control. Cell 81:323–330

Weirich-Schwaiger H, Weirich HG, Gruber B, Schweiger M, Hirsch-Kauffmann M (1994) Correlation between senescence and DNA repair in cells from young and old individuals and in premature aging syndromes. Mutat Res 316: 37–48

White E (1996) Life, death, and the pursuit of apoptosis. Genes Dev 10:1–15

Wilmut I, Schnieke AE, McWhir J, Kind AJ, Campbell KHS (1997) Viable offspring derived from fetal and adult mammalian cells. Nature 385:810–813

Wynford-Thomas D (1996) p53: guardian of cellular senescence. J Pathol 180:118–121

Yaagoubi AE, Mariethoz E, Jacquier-Sarlin MR, Polla BS (1998) Redox regulation of heat shock protein expression and protective effects against oxidative stress. In: Montagnier L, Olivier R, Pasqwer C (eds) Oxidative stress in cancer, AIDS and neurodegenerative diseases. Marcel Dekker, New York, pp 113–126

Zglinicki TV, Saretzki G, Docke W, Lotze C (1995) Mild hyperoxia shortens telomeres and inhibits proliferation of fibroblasts: a model for senescence? Exp Cell Res 220:186–193

Comparative Biology of Cell Immortalization

A. Macieira-Coelho[1]

1
Introduction

Higher metazoans are composed of different types of cells in regard to their long-term proliferative properties.

Gametes constitute a special compartment maintained in an haploid state, ready to be recruited to start a new organism. There is an optimum period during the organism life span for this recruitment, after which aging causes loss of this capacity. Cells such as neurons and skeletal muscle proliferate during embryonic development towards a terminal differentiation and become postmitotic. In some areas of the brain, however, neurons seem to keep a regenerating potential (Kempermann et al. 1997). Other cells such as hepatocytes are conditionally mitotic; in other words they are normally in a resting phase but can proliferate if needed to regenerate the liver. Other compartments are continuously regenerated through the life span of the organism. This is the case for the hemopoietic system, the gasterointestinal epithelium and the skin. The long-term capacity of these compartments for regeneration has not been ascertained. They are regenerated through the recruitment of stem cells, a reserve that seems to be present in most organs and which at least in some organisms have an unlimited proliferative capacity. But does this recruitment lead to cell lineages with an exhaustible mitotic potential?

The problem of the long-term proliferative potential of mitotic cells is not settled; some investigators claiming that normal cells have a finite potential (Hayflick 1965) and others explaining that when the latter is observed it is due to genetic damage (Rubin 1997). The pendulum has oscillated one way or the other according to the experimental system used and has not yet stood still.

It was hypothesized (Weissmann 1891) that somatic cells of higher organisms have evolved towards a finite proliferative potential due to an evolutionary trait. Weissmann proposed that when a functional character becomes unnecessary it eventually disappears. In the case of higher organisms, the finite division potential of somatic cells would cause the aging of

[1] University of Paris VI, 73bis rue Maréchal Foch, 78000 Versailles, France.

Progress in Molecular and Subcellular Biology, Vol. 24
A. Macieira-Coelho
© Springer-Verlag Berlin Heidelberg 1999

the soma thus creating the conditions for the elimination of the organism after the reproductive period. This would give an increased opportunity for new individuals to develop and improve the chance of survival of the species. Hence the implication of an association between the finite division potential of somatic cells and the mechanisms of aging.

Comparative biology, however, shows that when analysing the phenomenon in different kingdoms, in different species along the evolutionary scale, and in different cell types of a given species, the problem of the cell division potential looks much more complex. The behavior of cells in regard to this parameter, has some interesting bearings on the biology of the respective organisms, suggesting a relationship with evolutionary traits. It is difficult, however, to envision for the moment any kind of biological rule.

2
The Growth Potential of Cells from Different Origins

2.1
Proliferative Potential of Cells from Different Kingdoms and Different Phyla

In the plant kingdom cells seem to be endowed with an infinite growth potential. The cambium of some sequoia trees in California , which causes the trunk to grow in girth, has been growing with annual periods of rest or dormancy for 4,000 years (Steward 1963). The potential for growth of plant cells is also apparent from experiments showing that a new plant can be produced from nongrowing cells of the root provided adequate nutrients are present in the culture medium (Steward 1963). Moreover, tissues derived from a variety of plants have been maintained indefinitely in vitro, conserving for long periods the capacity to initiate a new plant (Torrey 1967). Eventually they lose this property, but are potentially immortal although they undergo genetic changes in culture (Phillips et al. 1994) which lead to the loss of the proliferative potential in some types of cells (Rubin 1997). These proliferative and developmental characteristics of plant cells are probably related to the great adaptability and reproductive potential of the plant kingdom, without which animal life would probably be impossible.

The situation is different in other kingdoms. The choice between finite or immortal proliferative phenotypes already becomes apparent in unicellular organisms. Bacteria, where growth constraints are dependent only upon the availability of nutrients, seem to possess only the immortal phenotype.

As other unicellular organisms evolved, the choice of finite or infinite life span became manifest. Clones of **Amoeba proteus** for instance, in an adequate nutrient environment are normally immortal. However, after a period of severe food restriction, cells subsequently transferred to normal feeding produce clones of finite life span (Danielli and Muggleton 1959). On the other hand, in cells from higher organisms (mouse fibroblasts) which have acquired an infinite growth potential, repeated rounds of prolonged growth constraint induce an enduring decrease in the rate of proliferation without the loss of the immortal phenotype (Rubin 1997).

The transfer of the nucleus from a food restricted amoeba at the end of its life span, into normal cytoplasm, produced a cell yielding a clone with limited life span (Muggleton and Danielli 1968). The same authors also reported that the limited life span could be induced in potentially immortal cells by the injection of cytoplasm from cells in decline. These results show that food restriction induces irreversible nuclear and cytoplasmic modifications determining the change in phenotype. One is tempted to wonder if this phenomenon is a way of eliminating organisms which have suffered too much damage during starvation.

The events in **Paramecium** are described in a separate chapter.

Sponges represent the lowest metazoan phylum. Their cells are immortal when aggregated and mortal when dissociated (Koziol et al. 1998).

Hydra is among the simplest multicellular animals. Cells having fulfilled their task are unceasingly replaced by substitutes generated from stem cells with an unlimited capacity to regenerate (Müller 1996). Under conditions of abundant food supply, cloned offsprings are produced by a process of budding.

In vivo studies suggest that in higher metazoans stem cells are endowed with infinite division potential. Experiments with the transplantation of imaginal disc blastemas of **Drosophila** larvae into adult flies demonstrated that these cells not only did not present any loss of the potential to proliferate, but also did not lose the capacity to differentiate (Hadorn 1969).

In the animal kingdom the mortal and immortal cell phenotypes have unravelled along the evolutionary scale, in certain species some cells have lost the immortal phenotype and others have kept it for reasons not yet understood.

Transplantation experiments with mouse hemopoietic cells have given variable results (Lajtha and Schofield 1971; Harrison 1984) some indicating a finite regenerating potential, while others suggest a possible immortal phenotype of bone marrow stem cells. These experiments, however, are difficult to interpret because they involve the total irradiation of the host, with possible adverse effects of the damaged environment on the transplanted cells. An increase in the interval between explants improves the regenerating potential of the stem cells (Vos and Dolmans 1972). This could mean that longer intervals allow for the regeneration either of the division potential of the transplanted cells through a resting period, or of the host tissues from the effects of the whole body irradiation. Experiments measuring the mitotic activity of mouse hemopoietic stem cells suggest that they do not lose growth potential with age and that mitotic activity is higher in old animals (Morrison et al. 1996).

It has been claimed that human hemopoietic stem cells show a loss of telomeric DNA with age, suggesting that their proliferative potential is limited (Vaziri et al. 1994). It has not been ascertained, however, if these are real stem cells.

Serially transplanted mouse mammary tissue which had lost the capacity to produce new buds, could regenerate new buds through stimulation of a layer of undifferentiated cap cells, the putative stem cell population with cholera toxin (Daniel et al. 1984).

In the ventral epithelium of the mouse tongue, experiments with DNA labelling showed that cells of the germinal layer displayed no signs of exhausting their long-term division potential (Cameron 1972). The same has been demonstrated for the stem cells of the mouse small intestine (Potten and Loeffler 1990).

In the mammalian forebrain the number of neural stem cells is maintained throughout life (Tropene et al. 1997).

In the liver it seems that two types of regeneration are possible: in the post-hepatectomy type of regeneration, mature hepatocytes are the cells which are recruited; when parenchymal cells are too damaged, the progenitor stem cells are those which regenerate the organ (Overturf et al. 1997). In both cases duct epithelial cells and differentiated hepatocytes are regenerated. Serial transplantation of adult mouse hepatocytes has not elucidated any limit to the capacity to regenerate the organ. Hence, in mouse liver it seems possible that not only stem cells but also differentiated hepatocytes are endowed with the immortal normal phenotype, allowing multipotency.

Experiments with cultivated cells also suggest that stem cells are endowed with an infinite proliferative potential without deviations from normalcy. Suda et al. (1987) could maintain mouse embryonic stem cells for more than 250 doublings. The cells were obtained before the division of soma and germ cells; they maintained the normal karyotype, could colonize embryos without causing developmental abnormalities, and were able to form normal gametes.

Mouse and human astrocyte precursors could also be maintained in vitro for at least 200 generations without any apparent deviations from normalcy (Loo et al. 1987, 1991).

So the data suggest that at least in some species, in lower and higher metazoans, the indefinite normal proliferative potential has been preserved in stem cells and maybe also in some other types of cells (e.g. hepatocytes). This seems to make sense from an homeostatic point of view. The situation is not clear in humans.

2.2
The Growth Potential of Fibroblasts from Different Species

In this chapter we will focus on fibroblasts because they have been extensively studied and because of the interesting relationships between the proliferative behavior of these cells and the physiopathologic and evolutionary characteristics of the respective organism.

Why should fibroblasts manifest this relationship? This has to be seen in the face of present knowledge concerning the role of mesenchymal-epithelial cell interactions in development and homeostasis. It should be remembered that mesenchymal cells are ubiquitous in the organism and have the task of creating a microenvironment through the secretion of soluble substances and of large molecules of the interstitial tissue. This property gives them a fundamental role during development through inductive interactions with other cell systems. Development of the thymus, for instance, depends on an interaction between mesenchyme and epithelium (Auerbach 1960) and mouse embryo mammary rudiments fail to develop in the absence of mesenchyme (Kratochwill 1969). These are just two examples of the functions of mesenchymal cells during development, for which there is an almost endless list of experimental evidence available. The influence of mesenchymal cells on other cell compartments proceeds in the mature organism and its disturbance can lead to disease (Elias et al. 1985). So the fibroblastic mesenchymal cell appears to be a universal regulator in the organism, and its properties reveal some fundamental characteristics of the milieu to which it belongs.

2.2.1
Species with Fibroblasts with an Immortal Phenotype

The cultivation of cells in vitro has greatly contributed to the debate as to whether somatic cells have conserved the finite or infinite proliferative potential phenotype.

Fibroblast-like cells derived from the caudal fin of the fish **Carassius auratus** could be maintained in vitro for more than 200 population doublings, corresponding to an observation period of 22 months (Shima et al. 1980). No significant change in the distribution of chromosome number was seen. The cells also did not manifest any signs of developing malignant properties. The authors suggested that the unlimited growth of these cells in vitro could be related to the continued growth of the living organism through its life span. This is an interesting hypothesis worth testing further.

Identical results were reported for marine teleost fish tissues (Sigel and Beasley 1973). The cells had undergone over 300 subcultures during an observation period of over 10 years without any alteration of morphology, growth rate or karyotypical status.

Embryos from **Rana pipiens** and other amphibians can also yield normal cell populations capable of propagating indefinitely in vitro (700 doublings) without any apparent modifications (Freed and Mezger-Freed 1973). One is tempted to relate this to the ability of amphibians to regenerate missing body parts.

Fibroblasts from some marsupial species, though not all, can be propagated indefinitely in vitro without any apparent deviations from normalcy (Stanley et al. 1975; Pye et al. 1977).

Hence it seems that in some orders all species have preserved the immortality of their somatic fibroblasts, while others diverged in regard to this trait.

2.2.2
Species with Fibroblasts with a Finite Proliferative Potential

The main criticism against experiments along these lines in vitro concerns the artificial conditions of the environment. The limited life span of cells in vitro could be an artefact of maladjustment to the radical change of the environment (Rubin 1997).

Certainly it is known that using, for instance, natural substrata instead of glass or plastic for cell attachment, improves the life span of cells. Such was the case with chicken fibroblasts which could be maintained for much longer periods on a collagen substratum (Gey et al. 1974). However, the number of divisions was difficult to evaluate under such culture conditions. Moreover, since the original explant was routinely transferred, the regeneration of the cultures could have been due to cells which remained quiescent in the explants. Anyway, in these experiments a growth decline became manifest after a certain time in vitro.

Improvements in the nutrient medium such as the addition of hormones (Macieira-Coelho 1966) are also known to prolong proliferation of human fibroblasts, but the cells eventually stop proliferating.

The fact that in vitro experiments with stem cells fit in vivo experiments, and that the data obtained with fibroblasts have a bearing on evolutionary and physiopathologic characteristics of the respective organisms, as is further described below, counter the criticism that the results are an artefact conditioned by the artificial environment.

Another piece of evidence favouring the reliability of the in vitro results concerning the expression of a mortal or immortal phenotype is the fact that the cells expressing the former, when they become permanent cell populations, manifest new properties (Macieira-Coelho and Azzarone 1988) and a progressively increased susceptibility to become malignant, either spontaneously or through the action of carcinogens (Freeman et al. 1975; Yang et al. 1992). In most cases the transition to a permanent cell line occurs after what is called a growth crisis, during which the mitotic activity is strikingly reduced.

Moreover, in humans, cells from donors at high risk of cancer and from some cancer patients, when cultivated in vitro, display a greater probability of spontaneous or induced immortalization with the acquisition of new properties.

Hence we feel that there are good arguments favouring the idea that the limited division potential expressed in vitro as well as the tendency to overcome this phenotype and to transform, represent intrinsic cell properties. Additional arguments are given below.

During this century, probably mainly due to Carrel's results that reported the maintenance of proliferating chicken fibroblasts in vitro for several years (Carrel 1912), normal proliferative cells were believed to be able to divide indefinitely, maintaining all their normal characteristics. The difference between normal and cancer cells would consist not in the division potential, but rather in a deviation from the homeostatic control of growth of the latter, with the acquisition of new characteristics.

The first descriptions of cell lines obtained from mammalian tissues which could be subcultivated indefinitely were made in the 1940s (Earle 1943; Gey et al. 1949). These lines, however, were found to be oncogenic (Earle 1943; Foley and Handler 1957) and so the search continued for the immortal cell population which could maintain normalcy in vitro (Puck et al. 1958). This view was challenged first by Swim and Parker (1957) who observed that the failure of human fibroblasts to proliferate indefinitely in standard culture media is a more general phenomenon than was usually recognized, and could not be attributed to the artificial environment. They suggested that mammalian cells that acquire the ability to proliferate in vitro for indefinite periods have a bearing on malignancy. They considered, however, the possibility that with improved media and sufficient persistence, permanent lines of morphologically unaltered fibroblasts could be derived from normal human tissues.

Hayflick and Moorhead (1961) completely opposed this concept. They observed the regular loss of cultivated human embryonic lung fibroblasts after approximately the same number of doublings without ever obtaining permanent cell lines, a phenomenon later observed by innumerous laboratories. Hayflick and Moorhead concluded that this is an intrinsic property of these cells which would be endowed with a limited potential to proliferate, an expression of aging at the cellular level. They related to malignancy the acquisition of an infinite capacity to divide.

Later Hayflick (1965) observed that during development of the organism, human fibroblasts lose their proliferative potential, reinforcing his hypothesis.

The difficulty in obtaining permanent cell lines with human fibroblastic cultures is not shared by similar cells from other species. The immortal cells, however, produced at different frequencies according to the species, progressively deviate from normalcy and eventually acquire cancer-like properties.

The study of this phenomenon in vitro has been a valuable tool in studying the progression of cellular changes leading to cancer. This was possible through the observation of how normal cells behave in vitro, which then allowed the understanding of deviations from normalcy, i.e. transformation. It has been a laborious task which has met with several misunderstandings. Indeed the word "transformation" has been misused and misinterpreted in cell biology.

Transformation is supposed to encompass the spontaneous or induced changes occurring in a cell population in vitro, having analogies with what occurs in vivo during the development of neoplastic growth. Hence those changes should follow a pattern analogous to the ones leading to cancer and express at least some of the steps allowing a cell population in vivo to escape homeostasis. These steps are complex and in vitro we can probably analyse just a fraction. Transformation consists in the progressive acquisition of growth autonomy, such that a cell population becomes less dependent upon the different mechanisms which restrain the growth of normal cells. During transformation a cell population goes through a series of steps whose succession is unpredictable and which can stop anywhere along the path to full autonomy. Transformation represents the opposite of what occurs when cell populations with a finite division potential proliferate in vitro and become increasingly sensitive to growth-inhibiting mechanisms (Macieira-Coelho et al. 1966).

Spontaneous or carcinogen-induced transformation can start with the loss of the social organization of the cell population during crowding, due to a lesser restraint of cell-cell overlapping and a decreased sensitivity to contact inhibition of movement and of growth (Macieira-Coelho 1967a). The growth requirements of transformed cells can change so that they become less dependent upon the serum supplement of the medium (Macieira-Coelho 1967b). The cells can become less dependent on attachment to a solid substratum and are able to proliferate in an agarose semisolid medium or even to grow in suspension in a liquid medium, i.e. they become completely anchorage independent. This has analogies to what happens in vivo when tumor cells get loose from the tissue where they originated and acquire the capacity to survive when transported in the peripheral blood. The capacity to invade tissues in vivo and to kill a host when injected into laboratory animals signals that the cell population has become tumorigenic. The acquisition of an unlimited potential to divide is one of the steps in the progression towards autonomy. This can be preceded by an extended, although finite, life span. (For a complete review of the transformation steps see Macieira-Coelho 1990).

When one compares fibroblasts from different species according to the probability of spontaneously yielding a population of transformed cells with unlimited growth potential, one finds a spectrum from very low to 100 % probability (Fig. 1). The latter is found with rat and mouse fibroblasts. In general murine fibroblasts tend to have a higher probability of yielding permanent cell lines than for instance human and bovine. Chicken fibroblasts are at the top of the scale because of their apparent stable finite potential to divide and of its interesting relationship with the mechanisms of transformation of these cells by viruses. This does not mean, however, that all avian cells behave that way.

Chicken fibroblasts seem extremely stable in regard to limited division potential when cultivated in vitro. In spite of the routine use of these cells in laboratories all over the world, to the best of our knowledge, there is only one reliable publication reporting the spontaneous establishment of an immortal chicken fibroblast population (Kaaden et al. 1982).

The same can be said for human fibroblasts. There is only one bona fide spontaneously obtained immortal human fibroblast population that originated from a patient with melanoma (Mukherji et al. 1984; McCormick and

Fig. 1. Scale indicating, from top to bottom, the increasing probability of cells from different origins escaping the mortal phenotype and acquiring the capacity for unlimited proliferation

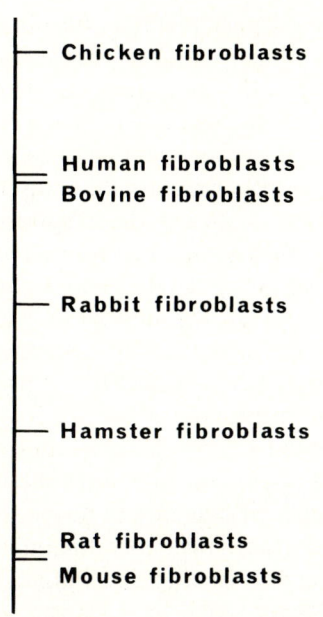

Maher 1988). This is an interesting finding with implications for the natural history of cancer which is discussed below. Another work reporting identical findings from four patients with lung tumors (Azzarone et al. 1976) was later found to be due to accidental laboratory contamination with hamster cells (A. Macieira-Coelho, unpubl. obs.).

Bovine fibroblasts also seem to be very stable in regard to their proliferative life span (Stenkvist 1966). They are, however, more susceptible to transformation by viruses than human cells.

For hamster cells there is a difference in regard to spontaneous immortalization, which depends upon the animal strain. Syrian hamster fibroblasts convert spontaneously to an immortal line with a frequency of 5 % (Bruce et al. 1986) to 10 % (Barrett 1980). Chinese hamster fibroblast cultures, however, from either fetal or postnatal origin, invariably become permanent cell lines (Kraemer et al. 1986).

Rat and mouse fibroblasts immortalize in vitro with a frequency of 100 % regardless of the animal strain (Freeman et al. 1975; Macieira-Coelho and Azzarone 1988).

Several results support the view that this phenomenon is not an artifact of in vitro subcultivation. One aspect which illustrates the relationship between the in vitro cell behavior and some characteristics of the respective organism, is the fact that it is easier to induce cancers in those species (e.g. murine) whose fibroblasts have a higher probability of acquiring an infinite division potential (Pontén 1971). Indeed anything can induce tumors in mice, even the implantation of pieces of plastic.

The species whose fibroblasts immortalize easily also seem to be more short-lived. Is there any relationship of cause and effect between cell instability and short life span? The question remains open but the relationship seems reasonable when one realises that the cellular instability seems to be related to some particularities of the genome which will be described below.

The other results showing that the in vitro proliferative potential is not an artifact, concern the response to carcinogens. Transformation of chicken fibroblasts, which are at one end of the scale, by Rous sarcoma virus (RSV) for instance, rarely yield a continuous cell line (Dinowitz 1977). In general the cells go through some transformation steps (e.g. piling up and loss of contact inhibition of growth) but do not immortalize and indeed have a shorter life span than the noninfected cells (Pontén 1971). Tumors developed in chicken after RSV infection also do not have a clonal type of growth (Pontén 1971). The spread of these tumors is due to increased cell proliferation after viral infection; the infected cells die after a few divisions and the virus is propagated to new cells which thus become recruited into the proliferative pool probably due inter alia to the release of autocrine growth factors (Macieira-Coelho and Pontén 1967; Macieira-Coelho et al. 1969).

Mouse cells, in contrast, are easily transformed by viruses and the cell population rapidly goes through many transformation steps with a clonal type of growth.

Human and bovine fibroblasts on the other hand, which have an intermediate position on the scale, when infected for instance with SV40 virus, go through some steps of transformation such as morphological changes, loss of contact inhibition of growth and prolonged life span, but stop short of full transformation since they in general phase out (Stenkvist 1966; Gotoh et al. 1979). Only rare clones can originate permanent cell lines. It has been claimed that a small-plaque SV40 variant can immortalize human fibroblasts with a 100 % efficiency (Miranda et al. 1988). This claim has not been repeated. The particular resistance of human fibroblasts to progress along the path of transformation and to immortalize, is manifested in response to all oncogenic viruses tested so far.

Identical responses can be obtained with chemical carcinogens; they easily make mouse cells to progress along different transformation steps (Mondal and Heidelberger 1977) but in general induce only some transformation steps with an extended life span in human fibroblasts without immortalization (Milo and Casto 1986).

The response to low dose rate ionizing radiation also fits the relative probability of the fibroblasts from the different species shown on the scale of Fig. 1 to spontaneously immortalize. Thus, this type of radiation shortens the life span of chicken embryo fibroblasts (Macieira-Coelho and Diatloff 1976) but it has no effect or can prolong the division potential of normal human embryonic fibroblasts (Macieira-Coelho et al. 1977; Croute et al. 1986), and it accelerates the immortalization of mouse fibroblasts (Macieira-Coelho and Diatloff 1976) (Fig. 2).

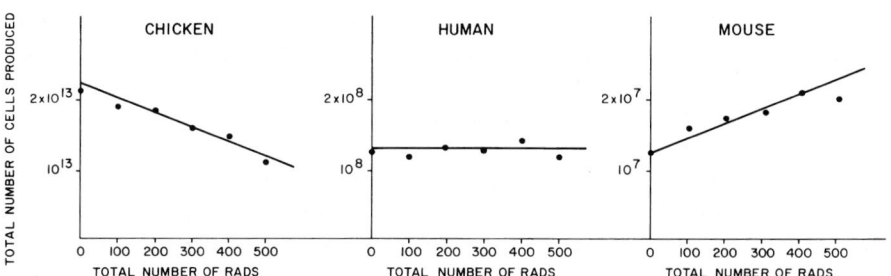

Fig. 2. Effect of different doses of low dose rate ionising radiation on the proliferative potential of cells from the scale illustrated in fig. 1. Each **point** represents the total number of cells produced after each dose of radiation, calculated by adding the increase in the number of cells after each subcultivation until the fibroblast population stoped proliferating. For mouse cells the points represent the total number of cells produced until the fibroblast population immortalized

These data suggest that the outcome of the action of all carcinogens on the fibroblasts of these different species, regardless of their nature (viral, chemical or physical) depends on the cell's latent potential to immortalize spontaneously.

Again, the effect of oncogenes is related to the relative position of these cells on the scale of Fig. 1. They, in general transform chicken cells without immortalization (Stehelin et al. 1976), in human fibroblasts they can induce some transformation steps such as focus formation and anchorage independence (Hurlin et al. 1987) but rarely immortalize them (Sager et al. 1983; Morgan et al. 1991), and they just accelerate the immortalization of murine fibroblasts (Land et al. 1983; Spandidos and Wilkie 1984). It is possible however, that there is a different propensity of human fibroblasts to immortalize according to body site. Human foreskin fibroblasts for instance seem to be more susceptible to transformation than those of other body sites (Kopelovich 1982). It has been suggested that they may represent a state in development that is genetically unstable and thus more susceptible to malignant transformation.

Additional results showing that the finite life span of human fibroblasts is not the result of an in vitro artifact report the presence of an increased number of terminal postmitotic fibroblasts in tissues in vivo in different pathological conditions (Macieira-Coelho 1995a).

2.2.3
Susceptibility of Fibroblasts from Different Human Donors to Immortalize

The interspecies diversity in the probability of escaping the mortal phenotype and to transform, has analogies with the diversity found in the susceptibility to immortalization of cells from different human donors.

It was mentioned above that the "normal" skin fibroblasts from a patient with melanoma immortalized spontaneously (Mukherji et al. 1984). Indeed, several works suggest that the fibroblasts from at least some cancer patients are unstable in regard to their proliferative potential and deviate from the criteria of normalcy.

This has interesting implications for the physiopathology of those cancers. There is clinical evidence in many cases of a field effect created by the mesenchyme, which is favorable for the development of tumor growth. In these cases, neoplastic disease has to be seen as a generalized disease where the whole organism is participating in its initiation. The mesenchyme would be important not only for the initiation of some neoplasiae but also for their maintenance and propagation since fibroblasts from the stroma of tumors have developed new properties relevant for interactions with the neoplastic cells that contribute to the propagation of the latter (Dabbous et al. 1987; Hamada et al. 1987).

It is not only the fibroblasts from cancer patients which are more prone to immortalize, those from donors with some genetic diseases which predispose to cancer also present the same tendency.

A higher rate of spontaneous immortalization has been reported with fibroblasts from patients with Li–Fraumeni syndrome (Bischoff et al. 1990). Permanent cell lines were also obtained with these cells after treatment with chemical carcinogens (Tsutsui et al. 1995) and X-ray irradiation (Tsutsui et al. 1997). The immortalization of human fibroblasts with chemical carcinogens has been obtained with cells from patients with von Recklinghausen's disease (Igel et al. 1975) which has a high rate of spontaneous transformation of neurofibromas to neurofibrosarcomas. Fibroblasts from some complementation groups of xeroderma pigmentosum patients have also been reported to be more susceptible to immortalization after infection with SV40 virus (Klein et al. 1990). Moreover the fibroblasts from patients with hereditary retinoblastoma are also more susceptible to immortalization by this virus (Banerjee et al. 1992).

3
Are Tumor Cell Populations Immortal?

It was mentioned above that immortalization is one of the steps along the path to the escape of growth control during the process of transformation. Cell populations that have immortalized are not inevitably capable of invading and killing an animal host, although in some cases the two modifications can be simultaneous. In most cases additional events are necessary for the cells to become malignant (Hei et al. 1997). Cells can also become malignant, for instance after treatment with chemical carcinogens, but still retain the mortal phenotype (Milo and DiPaolo 1978).

The susceptibility to transformation varies through the proliferative life span of cells with a finite division potential depending on the carcinogen (Macieira-Coelho 1994).

Tumor cells cultivated in vitro do not always yield immortal cell populations. Cells from colon cancer seem to readily develop into permanent cell lines (McBain et al. 1984); these results, however, cannot be extrapolated to all cancers. Cells from breast cancer metastases were obtained from one patient over a 2-year period, but despite repeated attempts, only the last specimen obtained at the end of that period reproducibly yielded cell populations with an infinite life span (Smith et al. 1987). Hence, in this case, cell immortalization was a late event in the progression of malignancy.

Attempts to develop permanent cell lines were also made with specimens obtained from the primary site and the peripheral lymph nodes from patients with small-cell lung cancer (Masuda et al. 1991). Twenty one percent permanent lines were obtained with specimens from the primary sites and 80 % from the peripheral lymph nodes. The survival times of the patients whose cells grew continually in culture were significantly shorter than those of the patients whose cells had a finite life span. These results suggest that the tendency of a tumor cell population to yield permanent cell lines is related to a more biologically aggressive form of the disease and have prognostic value (Masuda et al. 1991). Identical findings were obtained with neuroblastoma (Reynolds et al. 1980) and non-Hodgkin's lymphoma (Tweeddale et al. 1987).

The work of Masuda et al. (1991) also suggests that specimens obtained from metastases yield permanent cell lines more frequently and thus represent a more advanced stage of malignancy. This supports the findings of Smith et al. (1987) which suggest that immortalization, at least in some cancers, is a step along neoplastic progression.

It should be reminded that at least in some organisms the immortalization of the tumor cell is not an irreversible process since tumor cell populations from some organisms, in an appropriate environment, can reintegrate homeostasis and behave in a normal way (Steward 1963; Coleman et al. 1993).

On the other hand genome reshuffling with a nonspecific stimulus such as an electric pulse and an appropriate environment, the oocyte cytoplasm, can regenerate the growth potential of somatic cells and develop the whole organism (Wilmut et al. 1997).

4
Attempts to Identify the Mechanisms Involved in Cell Immortalization In Vitro

4.1
Nutritional Studies

One of the arguments against the belief that the mortal phenotype is a normal trait concerns the deficiency of the in vitro environment where the trait is expressed. On the contrary the in vitro conditions would be adequate for those cell populations that can divide indefinitely.

An experiment performed by Dell'Orco et al. (1973) strongly suggests that the nutrient media available are adequate for the maintenance of human fibroblasts. These investigators kept human fibroblast populations in medium supplemented with 0.5 % foetal calf serum. The cells can be kept under these conditions for several months with periodic medium renewals in a near resting stage; when resubcultivated and serially replicated in medium with 10 % serum, they will go through the same number of doublings as identical cultures that had been maintained constantly in a replicating state. This would not happen if the nutrient medium was deficient.

Adherents to the role of nurture have speculated that senescence of dividing cell compartments is an artifact of in vitro maintenance (Holliday et al. 1977); the immortal cells would be present in the primary culture but due to cultivation methods would not grow out. Cells committed to senescence would segregate at division of the immortal ones; subcultivation would dilute out the immortal cells so that after serial passages the population would be constituted only by cells committed to a finite life span. This hypothesis was claimed to be supported by a so-called bottleneck experiment where cells were plated at very low inocula at different population doubling levels (PDL); the procedure reduced the total number of doublings only when performed at early PDL and not at late stages of the cells' life span. These experiments were an artifact of cell maintenance since other investigators did not obtain the same results. Indeed maintaining the cells by serially plating them at low inocula yielded either a higher (Hayflick 1965) or the same number of doublings (Kaji and Matsuo 1980). We have also obtained a higher number of doublings maintaining the cells at a 1:16 rather than at a 1:2 split ratio (unpubl.).

Different types of experiments suggest that when cultivated cells yield immortalized populations with a high frequency (bottom of the scale of Fig. 1), the latter are not present in the original explanted specimen but develop through new properties acquired during subcultivation in vitro. Indeed, the extrapolation of the exponentially linear portion of the cumulative cell number to the time when the culture was initiated yielded a number smaller than the number of cells that initiated the culture (Todaro and Green 1963). Mouse fibroblasts during the period of decreasing proliferation (crisis) produce a colony-stimulating factor (Matsumura et al. 1979). When proliferation resumes after the growth crisis, the production of the stimulating factor decreases, indicating the development of new properties. Time lapse cinematography studies of the genealogies of clones of rat endothelial cells, reported in this volume by the same author, also suggest that the immortal cells originated in vitro.

The kinetics of cell proliferation analysed with bromodeoxyuridine (BrdU) labelling also point towards the in vitro origin of the immortal cells (Macieira-Coelho and Azzarone 1988). The experiments showed that in the primary culture, the dividing population was essentially composed of cells which performed three divisions during 48 h without any cells making only one division. The pool of rapidly dividing cells decreased when approaching the growth crisis, while at the same time a sharp increase in the fraction of slowly dividing cells was observed. When growth resumed, the slowly dividing cells (one division within 48 h) disappeared and the fraction of dividing cells was made of an identical percentage of cells which had performed two and three divisions within 48 h. When the fibroblasts immortalized, all mitoses corresponded to cells that had performed three divisions within 48 h. If the immortal cells were already present in the primary culture, one would expect that rapidly dividing cells increase progressively from the start.

Moreover, it was found that during immortalization of mouse fibroblastic cultures, the cells become progressively resistant to caffeine without the need for the cultures to be previously exposed to the drug (Macieira-Coelho and Azzarone 1988). Before immortalization, the cell population dies in the presence of caffeine; resistant cells start to appear only after the recovery from the growth crisis, and the development of resistance is progressive. If resistant cells were present from the beginning, in the presence of the drug they would outgrow the sensitive cells. Maintaining the cells in a larger volume of nutrient medium reduces the fall in cell number that characterizes the growth crisis. However, the time when the resistant cells become manifest is the same as when the growth crisis is more pronounced in the presence of a smaller volume of medium. These experiments show that caffeine resistance is not a selection phenomenon but is rather a new, spontaneously acquired property occurring in parallel with immortalization.

Not surprisingly there are nutritional differences between normal human cells and those prone to immortalization. A nutritional difference has been reported between normal postnatal fibroblasts and cells from patients with Gardner's syndrome and familial colon cancer; the growth of the latter cells was more sensitive to methionine deprivation (Mikol and Lipkin 1984). Other results showed that fibroblasts (susceptible to immortalization) from patients with some genetic diseases prone to cancer, have relaxed growth requirements and prolonged life span (Brothman et al. 1986).

4.2
Genetic Studies

Several experiments suggest possible explanations for the different susceptibility of cells in vitro to acquire an infinite proliferative life span; they concern observations at the cytogenetic level during serial divisions of these cell populations. These data suggest that the long-term doubling potential of cells is directly related to the potential for continuous chromosomal rearrangements, i.e. the plasticity of the genome.

Human embryonic fibroblasts when cultivated in vitro go through continuous chromosomal rearrangements without any definite pattern becoming apparent (Chen and Ruddle 1974; Harnden et al. 1976). The data showed that multiple clones arise continuously and compete between each other without any one overgrowing the others. These cells have a longer division potential than postnatal fibroblasts which go through more stable type chromosomal rearrangements during serial divisions in vitro (Harnden et al. 1976). Moreover, fibroblasts from Werner's syndrome patients, which have a reduced doubling potential when compared with cells from age-matched normal donors, present chromosomal rearrangements (variegated translocation mosaicism) which become fixed and remain predominant during the cell population in vitro life span (Salk et al. 1981). These findings suggest that a "rigidity" of the genome is associated with a shorter doubling potential as opposed to a higher "plasticity" which would favor a longer potential.

Another finding suggesting a direct relationship between limited division potential and loss of genome plasticity, is the one showing that in human fibroblasts close to the end of their proliferative life span, it becomes more difficult to induce sister chromatid exchanges (Schneider and Gilman 1979). These results suggest that the loss of the division potential is associated with a loss of the potential for recombination of the genome.

The response of fibroblasts to low dose rate ionizing radiation also suggests a causal relationship between the capacity of the genome to recombine and the probability of escaping a finite proliferative life span. When human embryonic fibroblasts were treated with low dose rate radiation at low population doubling level, their life span could be prolonged (Macieira-Coelho et al. 1977); but if irradiated late during their life span, when the potential number of chromosome rearrangements was reduced (Schneider and Gilman 1979) their doubling potential was shortened (Macieira-Coelho et al. 1977). The same type of radiation shortens the division potential of early passage postnatal fibroblasts whose genome has a lower potential for chromosomal recombinations than that of embryonic cells (Bourgeois et al. 1981). On the other hand, the same treatment can significantly prolong the life span of human postnatal fibroblasts from donors with genetic diseases with a high rate of chromosome recombinations, which are at high risk of cancer (Macieira-Coelho 1994). In the irradiated cells, most of the breaks involved in exchanges concerned the centromeric and telomeric regions which are rich in repetitive DNA and have been implicated in recombinational events.

Furthermore, the skin fibroblasts of melanoma patients, which are susceptible to immortalization, have an increased UV-induced sister chromatid exchange (Knees–Matzen et al. 1991). It was also found (Macieira-Coelho 1994) that the prolongation of the doubling potential of skin fibroblasts from retinoblastoma patients by low dose ionizing radiation showed a correlation with the potential for sister chromatid exchanges of the cells of the respective patients. Hence, the presence of the deletion was not enough to confer susceptibility to the carcinogen; other genetic factors, which confer to the genome a higher recombination potential, determine the proneness of these cells to transformation by radiation (Macieira-Coelho 1994).

Cells from Bloom's syndrome patients, an inherited disease with a high risk of developing cancer, exhibit elevated levels of somatic crossing-overs and sister chromatid exchanges (German 1993). A gene was identified (BLM) which encodes a protein homologous to the one coding for a helicase subfamily first identified in bacteria and yeast (Rothstein and Gangloff 1995). Mutations in BLM and in the yeast homologue confer the common phenotype of hyper-recombination to cells.

Moreover, a protein has been isolated from malignant cells which seems important in the induction of chromosomal translocations (Chalk et al. 1997); the elevated recombination in immortal human cells was found to be mediated by a recombinase (Xia et al. 1997).

The positive correlation between chromosome rearrangements and the cell long-term growth potential fits the hypothesis proposed by Boveri at the start of the century which established a relationship between such rearrangements and malignant transformation. It also agrees with the proposal that malignant transformation results from recombination in cellular genes rather than mutations in latent cancer genes (proto-oncogenes; Duesberg 1987).

On the other hand, the mouse genome seems to be endowed with the capacity to survive pronounced chromosome recombinations and reorganizations. Indeed, the chromosomes of mouse fibroblasts cultivated in vitro display a very unusual capacity for recombination which is expressed by the presence of crossing overs and bridges between chromosomes (Macieira-Coelho and Azzarone 1988).

Another difference between the mouse and the human genome is the higher rate of sister chromatid exchanges in the former and the rapidity with which mouse cells can switch from the diploid to the tetraploid state (Macieira-Coelho and Azzarone 1988). Furthermore, the chromatin lability, which accumulates during cellular aging, disappears in the case of mouse cells during these chromosomal rearrangements which occur during the transition to immortalization (Macieira-Coelho 1991). DNA elimination in mouse fibroblasts during the period preceding immortalization, could be germane to the disappearance of the fragile chromatin sites. DNA measurements on cells in interphase after ethidium bromide staining showed that occasionally at early PDL, the DNA content of G2 cells is less than that expected from the G1 content (Macieira-Coelho and Azzarone 1988). This suggests that a process of DNA elimination goes on in these cells as has been reported for lower organisms (Yao et al. 1984). Although DNA elimination also takes place during the division cycle of human cells, it is not of the same order of magnitude as in mouse cells (Macieira-Coelho et al. 1982).

The potential for chromosomal recombinational events expressed by mouse fibroblasts in vitro also has a counterpart in vivo. Although 40 acrocentric chromosomes is the usual diploid number of the mouse species, localized races with 38 to 22 chromosomes resulting from Robertsonian fusions, have been found in the wild (Capanna 1973). This property of the mouse genome could be responsible for the high probability of mouse cells to spontaneously immortalize and acquire malignant characteristics, for the high susceptibility to viral, chemical and physical carcinogens and to onco-genes, and for the facility with which one can induce tumors in mice.

Attempts were made to identify specific genes responsible for the mortal phenotype, introducing human chromosomes into immortal cell lines. It soon became apparent that several chromosomes can suppress the immortal phenotype and that multiple genes are lost or mutated in immortal cell lines (Tsutsui et al. 1997). One of the problems involved with this type of experi-ment concerns the use of A9 mouse fibroblasts to select the human chromo-somes. It is known that there is cross contamination of the two genomes in hybrids between human and rodent cells (Littlejohn et al. 1995). Fusion products between human and mouse cells showed that species-related genetic determinants may interfere with the expression of certain pheno-types rendering impossible an interpretation of the results according to Mendelian genetics (Kaouël et al. 1978). It is an interesting aspect that was for the first time reported in this work, and has been overlooked in the inter-pretation of other experiments along these lines.

Fusions between normal and immortalized cells were used to study the genetics of the escape from the mortal phenotype and acquisition of an infi-nite growth potential. Muggleton-Harris and DeSimone (1980) studied the replicative potentials of fusion products between normal WI-38 cells and identical cultures transformed by SV40. The replicative potential of the hybrids was similar to the control WI-38 cells in the majority of cases. Only less than 2 % of the hybrids had a prolonged life span and possibly immor-talized. Fusions between normal WI-38 cytoplasts and transformed karyo-plasts did not show sustained replication. Mass culture hybridizations between Lesh-Nyhan cells with finite life span (deficient in hypoxanthin-phosphoribosyl transferase) and HeLa cells showed that most hybrids had a limited life span slightly longer than Lesh-Nyhan fibroblasts; only 1 in 10^5 cells were capable of infinite replication (Bunn and Tarrant 1980). Fusions of Lesh-Nyhan cells with HeLa cytoplasts did not show transformation.

The division potential of clones resulting from hybridizations between mortal and immortal human cell lines was analysed (Pereira-Smith and Smith 1983). Since the majority of clones had a finite life span it was con-cluded that cellular immortality is recessive although a fraction of clones yielded immortal cell populations. This fraction, although significant, was not taken into consideration; it was argued that certain cell types are more unstable in hybrids and yield immortal variants more rapidly (Pereira-Smith et al. 1990). Four complementation groups (A,B,C and D) were assigned for indefinite division to explain the results obtained with crosses between immortal cell lines (Pereira-Smith and Smith 1988). Further studies along these lines showed that four complementation groups were not enough to explain the data and that a further subdivision of the immortal cell lines into multiple complementation groups A, B, C, D, E ...., was necessary (Moy et al.

1997). The difficulty in the interpretation of these experiments was buttressed by other works showing that there are more chromosomes identified by monochromosome transfer studies containing a putative gene responsible for the mortal phenotype than there are complementation groups (Vojta and Barrett 1995).

Inter alia, these results have to be interpreted keeping in mind the data obtained by Martinez et al. (1978), showing that clones from the HeLa tumor derived cell line display a nonproliferative and a proliferative pool. The fastest growing clones and subclones segregate daughter cells with different growth potentials, showing a shift in the distribution toward lower growth rates and increase in the proportion of cells belonging to the nonproliferative pool. This phenomenon was called clonal attenuation. The same phenomenon exists in normal human fibroblast populations, but whereas the proliferating fraction disappears progressively in these cultures, in HeLa cells the two distinct pools are continuously produced. Hence in the fusions described above between immortalized and normal cells, some of the hybrids may have been obtained from cells committed to clonal attenuation. If HeLa cells display clonal attenuation, all other immortalized cell populations must have it to different degrees. It is possible that the different life spans obtained with the hybrids, depending on the type of immortalized parental cell line, depends on the degree of clonal attenuation displayed by the latter. On the other hand, the methods used for fusion may cause genetic changes which accentuate the phenomenon of clonal attenuation. Another publication shows that this may indeed occur. Evidence was obtained showing that the use of drug selection to obtain the hybrids is deleterious and that the arrest of division may result from the cytotoxic effect of drug selection rather than from the expression of the mortal phenotype (Ryan et al. 1994).

An additional flaw is related to the definition of the postmitotic state in the hybrids with finite life span. The postmitotic terminal cell of a normal human fibroblast population is a well defined state with identifiable markers whose presence was rarely if at all checked in the mortal hybrids. The growth arrest may have been due to nonspecific genetic damage.

Moreover the product of fusion between two nuclei is unpredictable and during serial divisions the remodelling going on in the genomes is random. This renders the outcome uninterpretable. Immortalization is a complex genetic phenomenon which cannot be due to just a few genes and cannot be interpreted in terms of Mendelian genetics.

Finally, still concerning putative genetic causes of cell immortalization, although a chapter in this volume describes the subject of telomeres, we will review here some aspects of the relationship between changes in telomeres and the cell's proliferative life span, which are relevant for the comparative biology of cell immortalization.

It has been postulated that the inability of DNA polymerase to fully replicate the ends of the discontinuously synthesized DNA strand leads to a shortening of the chromosome ends at each cell division cycle (Olovnikov 1973). The loss to a critical level of TTAGGG repeats that characterize mammalian chromosome ends, would cause the arrest of cells.

The hybridization of the terminal restriction fragments (TRF) of the DNA from serially dividing human fibroblasts with the appropriate probe, showed a shortening of the hybridization signal which was interpreted as a shortening of the telomeres with increasing PDL (Harley et al. 1990). The extension of the human fibroblast life span accompanied by telomere elongation after the introduction of telomerase into the cells, supports the relationship of causality between telomere shortening and the cells' finite division potential (Bodnar et al. 1998).

Furthermore, since some immortal cell lines express the enzyme telomerase and develop the capacity to reconstitute telomeres after each replication, the link between telomere integrity and replication potential seemed established (Harley 1991). Several investigators quickly jumped to the conclusion that the key to malignancy was the acquisition of the ability to regenerate chromosome ends. However, screening of different kinds of tumors for the presence of the canonical human telomeric repeat $(TTAGGG)_n$ gave variable results. In some studies 40 % of the tumors analysed revealed a reduction of telomere repeat arrays (Schmitt et al. 1994). Moreover, it was reported that telomere repeat fragment size does not limit the growth potential of leiomyomas (Rogalla et al. 1995). Exceptions to the rule also occur in normal human cells. Thus exhaustion of the growth potential of normal human oral keratinocytes is associated with loss of telomerase activity without shortening of telomeres (Kang et al. 1998).

On the other hand, fibroblasts from patients with Werner's syndrome, which have a lower proliferative potential compared with those of normal age-matched control donors, do not have shorter telomeres at low PDL than those of the control cells (Schulz et al. 1996). The mean TRF lengths of Werner's cells which had ceased replication were significantly longer than those of the controls.

The situation is also not identical in all species. Mouse fibroblasts which have a shorter proliferative life span than human fibroblasts, have chromosomal terminal repeats that are many times longer (Kipling and Cook 1990). Their size is also largely unchanged through somatic cell division. Mouse TRF is highly polymorphic, suggesting an unusually high mutation rate (Kipling and Cook 1990). This could be another marker expressing the high plasticity of the mouse genome mentioned above. Moreover, mice without telomerase obtained with gene knock-out techniques, reproduce normally and their cells are as efficiently immortalized as those from telomerase positive animals (Blasco et al. 1997).

Other aspects that have to be explained for the understanding of the implications of changes at chromosome ends for proliferative life span, depends upon the clarification of what really happens at the level of telomeres during replication. Approximately a quarter of in vitro immortalized cell lines have no detectable telomerase activity but have long telomeres (Bryan and Reddel 1997). Hence they must have an alternative mechanism for chromosome lengthening. Yeast apparently uses recombination and Drosophila uses specific retrotransposons for this purpose (Biessmann and Mason 1997).

The telomeres of the chromosomes of human fibroblasts were analysed with the canonical probe which detects all telomeres and with a TelBam11 probe which is specific for a subset of human telomeres (Ben 1997). A reduction in size of TRFs was found with the former probe; there was no evidence however, for loss of TelBam11 homologous sequences even in cells at the end of their life span. The heterogeneity in fragment size was much greater when the canonical probe was used, compared with the TelBam11 probe. This could be attributable to greater variability in the location of the restriction enzyme sites rather than differences in the length of terminal repeats. Ben concluded that variation in terminal repeat length may be related to the extent that telomeres participate in chromosome rearrangement. Telomere-promoted recombination can lead to degeneration of the telomeric sequence and subsequent loss of the hybridization ability (Ashley and Ward 1993). Indeed several examples of apparent terminal deletions are in fact subtelomeric translocations (Meltzer et al. 1993). Since TTAGGG repeats seem particularly prone to recombinational events, the events taking place at chromosome ends could be part of the overall reorganization of the genome during proliferation which eventually contributes to a postmitotic state. Some species for as yet unknown reasons may be more able to overcome the reorganization and immortalize.

4.3
Other Parameters which Could Influence the Probability to Escape the Mortal Phenotype

The presence of C-type virus particles in rodent cells was thought to act as an activator for the acquisition of the immortal phenotype (Freeman et al. 1975). However, identical particles are known to be present in normal human fibroblasts (Panem et al. 1975).

Several interspecies differences are known which could be implicated in the acquisition of an immortal phenotype, either spontaneously or in response to transforming agents. The capability of turning a carcinogen into a mutagenic form and of metabolizing the carcinogen into a water soluble form differs in cells from species with different life spans and could be implicated in susceptibility to transformation (Schwartz and Moore 1977).

Another important difference concerns DNA repair. Both the initial rate and the maximum incorporation of 3H-thymidine after UV irradiation was found to be higher in human and bovine than in hamster, rat and mouse cells (Hart and Setlow 1976). Another type of repair where interspecies differences have been described is the removal of **Micrococus luteus** UV-endonuclease susceptible sites. Human fibroblasts maintain this type of repair, in contrast to rat cells which lose it during serial proliferation (Mullaart et al. 1988). These authors suggested that this characteristic may render rat cells more dependent on postreplication repair systems which are error prone. It may lead to an increased frequency of genetic changes in rat cells, which could have a bearing on their susceptibility to transformation described above. On the other hand, mouse cells have a reduced capacity for excision-repair as revealed by the low host–cell reactivation of UV-irradiated herpes simplex virus (Yagi 1982; Elliot and Johnson 1985). Single strand break repair also differs, the relative increased efficiency of this type of repair is mouse<chicken<human fibroblasts (Diatloff-Zito et al. 1981). The repair of potential lethal damage is also more efficient in human and chicken than in murine cells (Diatloff-Zito et al. 1981). DNA repair mechanisms have been studied in almost all human genetically determined diseases prone to cancer after the discovery of the defective repair in xeroderma pigmentosum cells (Cleaver 1968) and different types of repair defects have been described. DNA repair, however, although unquestionably important for the transformation process, must be only part of the picture.

The mechanisms responsible for an increased susceptibility of fibroblasts from cancer prone individuals and from cancer patients without any known genetic predisposition to cancer to acquire an immortal phenotype is also a complex question involving multiple parameters. In general, there is a genetic instability expressed as an increased level of spontaneous and mutagen-induced chromosome aberrations (Heim et al. 1985). An interesting finding concerned the presence of a higher level of DNA transcripts from a subgenomic fraction obtained from Fanconi's anemia and retinoblastoma skin fibroblasts, which is also preferentially expressed in all human tumor cells tested so far (Hanania et al. 1988).

Studies with the fibroblasts from some cancer patients identified a subpopulation of cells with distinctive properties which precede the manifestation of the disease (Azzarone et al. 1984). The cells have less stringent conditions for growth concerning the interaction with the substratum and with other cells, displaying abnormal invasive properties. They also have a migratory foetal-like behavior that suggests that these patients may have a developmental defect resulting in a left-over of an embryonic phenotype more susceptible to transformation (Schor et al. 1988). One of the features characterizing this trait concerns the production of a migration-stimulating factor produced by embryonic but not by postnatal cells. Moreover, while normal embryonic cells stop producing this factor at the end of their life span, the fibroblasts from breast cancer patients do not undergo such a phenotypic transition. The persistence of this phenotype in the fibroblasts of the cancer patients may perturb the normal epithelial-mesenchymal interactions, creating a favorable terrain for the malignant transformation of the breast epithelium (Schor et al. 1989).

The increased migratory potential of the abnormal fibroblasts is related to the increased motility and reduced cell-to-substrate adhesion of the fibroblasts from cancer prone patients, associated with a defect in actin microfilament organization (Higgins and Kopelovich 1991) and reduced half-life of actin (Antecol et al. 1986). Mutant beta-actin expression and tropomyosin isoform switching are known to accompany immortalization of some fibroblasts (Varma and Leavitt 1988).

Cells prone to immortalization also have a defect in the G2 period of the division cycle leading to an increased transit time during this period (Azzarone and Macieira-Coelho 1987). This was attributed to a defective G2 checkpoint involving the inhibition of $p34^{cdc2}$/cyclin B protein kinase activity (Paules et al. 1995), a necessary step in the prevention of genetic damage through DNA excision repair (Parshad et al. 1983).

Finally, from the data reviewed above it is obvious that the phenomenon of escaping the mortal phenotype and immortalizing is extremely complex, involving innumerous different parameters. Simple models proposing two mortality stages, M1 and M2, regulated by specific genes that have to be overcome for cell immortalization to take place (Shay et al. 1991), are incompatible with the complexity of events leading to the mortal cell phenotype (Macieira-Coelho 1995b).

References

Antecol MH, Darveau A, Sonenberg N, Mukherjee BB (1986) Altered biochemical properties of actin in normal skin fibroblasts from individuals predisposed to dominantly inherited cancers. Cancer Res 46:1867–1873

Ashley T, Ward DC (1993) A "hot spot" of recombination coincides with an interstitial telomeric sequence in the Armenian hamster. Cytogenet Cell Genet 62:169–176

Auerbach R (1960) Morphogenetic interactions in the development of the mouse thymus gland. Dev Biol 2:271–282

Azzarone B, Macieira-Coelho A (1987) Further characterization of the defects of skin fibroblasts from cancer patients. J Cell Sci 87:155–162

Azzarone B, Pedull... D, Romanzi CA (1976) Spontaneous transformation of human skin fibroblasts derived from neoplastic patients. Nature 262:74–75

Azzarone B, Mareel M, Billard C, Scemama P, Chaponnier C, Macieira-Coelho A (1984) Abnormal properties of skin fibroblasts from patients with breast cancer. Int J Cancer 33:759–766

Banerjee A, Srivatsan E, Hashimoto T, Takahashi R, Xu H-J, Hu SX, Benedict WF (1992) Immortalization of fibroblasts from two patients with hereditary retinoblastoma. Anticancer Res 12:1347–1354

Barret JC (1980) A preneoplastic stage in the spontaneous neoplastic transformation of Syrian hamster embryo cells in culture. Cancer Res 40:91–94

Ben P (1997) Aging chromosome telomeres: parallel studies with terminal repeat and telomere associated DNA probes. Mech Ageing Dev 99:153–166

Biessmann H, Mason JM (1997) Telomere maintenance without telomerase. Chromosoma 106:63–69

Bischoff FZ, Yim SO, Pathak S, Grant G, Siciliano MJ, Giovanella BC, Strong LC, Tainsky MA (1990) Spontaneous abnormalities in normal fibroblasts from patients with Li-Fraumeni cancer syndrome: aneuploidy and immortalization. Cancer Res 50: 7979–7984

Blasco MA, Lee H-W, Hande MP, Smaper E, Lansdorp PM, DePinho RA, Greider CW (1997) Telomere shortening and tumor formation by mouse cells lacking telomerase RNA. Cell 91:25–34

Bodnar AG, Ouellette M, Frolkis M, Holt SE, Chiu C-P, Morin GB, Harley CB, Shay JW, Lichtsteiner S, Wright WE (1998) Extension of life span by introduction of telomerase into normal human cells. Science 279:349–352

Bourgeois CA, Raynaud N, Diatloff-Zito C, Macieira-Coelho A (1981) Effect of low dose rate ionizing radiation on the division potential of cells in vitro. VIII. Cytogenetic analysis of human fibroblasts. Mech Ageing Dev 17:225–231

Brothman AR, Cram LS, Bartholdi MF, Kraemer PM (1986) Preneoplastic phenotype and chromosome changes of cultured human Bloom syndrome fibroblasts. Cancer Res 46:7191–7197

Bruce SA, Deamond SF, Ts'o (1986) In vitro senescence of Syrian hamster mesenchymal cells of fetal to aged origin. Inverse relationsip between in vivo donor age and in vitro proliferative capacity. Mech Ageing Dev 34:151–173

Bryan TM, Reddel RR (1997) Telomere dynamics and telomerase activity in in vitro immortalized human cells. Eur J Cancer 33:767–773

Bunn CL, Tarrant GM (1980) Limited lifespan in somatic cell hybrids and cybrids. Exp Cell Res 127:385–394

Cameron IL (1972) Minimum number of cell doublings in an epithelial cell population during the lifespan of the mouse. J Gerontol 27:157–161

Capanna E (1973) Cytotaxonomy and vertebrate evolution. Academic Press, London

Carrel A (1912) On the permanent life of tissues outside the organism. J Exp Med 15:516–527

Chalk JG, Barr FG, Mitchell CD (1997) Translin recognition site sequences flank chromosome translocation breakpoints in alveolar rhabdomyosarcoma cell lines. Oncogene 15:1199–1205

Chen TR, Ruddle FH (1974) Chromosome changes revealed by the Q-band staining method during cell senescence of WI-38. Proc Soc Exp Biol Med 147:533–537

Cleaver JE (1968) Defective repair replication of DNA in xeroderma pigmentosum. Nature 218:652–654

Coleman WB, Wennerberg AE, Smith GJ, Grisham JW (1993) Regulation of the differentiation of diploid and some aneuploid rat liver epithelial (stemlike) cells by the hepatic environment. Am J Pathol 142:1372–1382

Croute F, Vidal S, Soleilhavoup JP, Vincent C, Serre G, Planel H (1986) Effects of a very low dose rate of ionizing radiation on the divison potential of human embryonic lung fibroblasts in vitro. Exp Gerontol 21:1–7

Dabbous MK, Haney, Carter LM, Paul AK, Reger J (1987) Heterogeneity of fibroblast response in host-tumor cell-cell interactions in metastatic tumors. J Cell Biochem 35:333–338

Daniel CW, Silberstein GB, Strickland P (1984) Reinitiation of growth in senescent mouse mammary epithelium in response to cholera toxin. Science 224:1245–1247

Danielli JF, Muggleton AL (1959) Some alternative states of amoeba, with special reference to life span. Gerontologia 3:76–90

Dell'Orco RT, Mertens JG, Kruse PF Jr (1973) Doubling potential, calendar time, and senscence of human diploid cells in culture. Exp Cell Res 77:356–362

Diatloff-Zito C, Deschavanne PJ, Loria E, Malaise E, Macieira-Coelho A (1981) Comparison between the radiosensitivity of human, mouse and chicken fibroblast-like cells using short-term endpoints. Int J Radiat Biol 39:419–426

Dinowitz M (1977) A continuous line of Rous sarcoma virus-transformed chick embryo cells. J Natl Cancer Inst 58:307–312

Duesberg P (1987) Cancer genes: rare recombinants instead of activated oncogenes. A review. Proc Natl Acad Sci USA 84:2117–2121

Earle WR (1943) Production of malignancy in vitro IV. The mouse fibroblasts, cultures and changes seen in the living cells. J Natl Cancer Inst 4:165–212

Ejila Y, Oshimura M, Sasaki MS (1990) Establishment of a novel immortalized cell line from ataxia telangiectasia fibroblasts and its use for the chromosomal assignement of radiosensitvity gene. Int J Radiat Biol 58:989–997

Elias JA, Zurier RB, Schreiber AD, Leff JA, Danièle RP (1985) Monocyte inhibition of lung fibroblast growth: relationship to fibroblast prostaglandin production and density defined monocyte subpopulations. J Leukocyte Biol 37:15–24

Elliot GC, Johnson RJ (1985) DNA repair in mouse fibroblasts. II. Responses of nontransformed preneoplastic and tumorigenic cells to ultraviolet irradiation. Mutat Res 145:185–190

Foley GE, Handler AH (1957) Differentiation of normal and neoplastic cells maintained in tissue culture by implantation into normal hamsters. Proc Soc Exp Biol Med 94:661–664

Freed JJ, Metzger-Freed L (1973) Frog embryos (haploid lines). In: Kruse PF Jr, Patterson MK (eds) Tissue culture, methods and applications. Academic Press, New York, pp 123–128

Freeman AE, Igel HJ, Price PJ (1975) I. In vitro transformation of rat embryo cells: correlations with the known tumorigenic activities of chemicals in rodents. In Vitro 11:107–116

German J (1993) Bloom syndrome: a mendelian prototype of somatic mutational disease. Medicine 72:393–404

Gey GO, Gey MK, Firor WM, Self WO (1949) Cultural and cytological studies on autologous normal and malignant cells of specific in vitro origin. Acta Unio Int Cancer 6:706–712

Gey GO, Svotelis M, Foard M, Bang FB (1974) Long-term growth of chicken fibroblasts on a collagen substrate. Exp Cell Res 84:63–71

Gotoh S, Gelb L, Schlessinger D (1979) SV40-transformed human diploid cells that remain transformed throughout their limited life span. J Gen Virol 42:409–415

Hadorn E (1969) Proliferation and dynamics of cell heredity in blastema cultures of Drosophila. Natl Cancer Inst Monogr 31:351–364

Hamada JI, Takeichi N, Kobayashi H (1987) Inverse correlation between the metastatic capacity of cell clones derived from a rat mammary carcinoma and their intercellular communication with normal fibroblasts. Jpn J Cancer Res (Gann) 78: 1175–1179

Hanania N, Diatloff-Zito C, Schaool (1988) An abnormal expression of a tumor-activated multigenic set in cells from cancer prone patients with inherited Fanconi's anemia and retinoblastoma. Cancer Lett 39:29–31

Harley CB (1991) Telomere loss: mitotic clock or genetic time bomb. Mutat Res 256:271–283

Harley CB, Futcher AB, Greider CW (1990) Telomeres shorten during ageing of human fibroblasts. Nature 345:458–460

Harnden DG, Benn PA, Oxford JM, Taylor AMR, Webb TP (1976) Cytogenetically marked clones in human fibroblasts cultured from normal subjects. Somatic Cell Genet 2:55–60

Harrison DE (1984) Do hemopoietic stem cells age? Cell Ageing Monogr Dev Biol 17:21–41

Hart RW, Setlow RB (1976) Correlation beween deoxyribonucleic acid excision-repair and life span in a number of mammalian species. Proc Natl Acad Sci USA 5:67–71

Hayflick L (1965) The limited in vitro life time of human diploid cell strains. Exp Cell Res 37:614–636

Hayflick L, Moorhead PS (1961) The serial cultivation of human diploid cell strains. Exp Cell Res 25:585–621

Hei TK, Wu LJ, Piao CQ (1997) Malignant transformation of immortalized human bronchial cells by asbestos. Environ Health Perspect 105:1058–1088

Heim S, Johansen SG, Kolnig AM, Strömbeck B (1985) Increased levels of spontaneous and mutagen-induced chromosome aberrations in skin fibroblasts from patients with adenomatosis of the colon and rectum. Cancer Genet Cytogenet 17:333–346

Higgins PJ, Kopelovich L (1991) Analysis of actin microfilaments and cell-to-substrate adhesive structures in human fibroblasts from individuals genetically predisposed to colonic carcinoma. Exp Cell Res 195:395–400

Holliday R, Huschtscha LI, Tarrant GM, Kirkwood TBL (1977) Testing the commitment theory of cellular aging. Science 198:366–370

Hurlin PJ, Fry DG, Maher VM, McCormick JJ (1987) Morphological transformation, focus formation, and anchorage independence induced in diploid human fibroblasts by expression of a transfected H-ras oncogene. Cancer Res 47:5752–5758

Igel HJ, Freeman AE, Spiewak JE, Kleinfeld KL (1975) Chemical transformation of diploid human cell cultures: a rare event. In Vitro 11:117–129

Kaaden O-R, Lange S, Stiburek B (1982) Establishment and characterization of chicken embryo fibroblast clone LSCC-H32. In Vitro 18:827–834

Kaji K, Matsuo M (1980) A low density inoculation method for the serial subcultivation of human diploid fibroblasts: an efficient model system for the study of cellular aging. Mech Ageing Dev 13:219–224

Kang MK, Guo W, Park N-H (1998) Replicative senescence of normal human oral keratinocytes is associated with the loss of telomerase activity without shortening of telomeres. Cell Growth Differ 9:85–95

Kang MK, Guo W, Park N-H (1998) Replicative senescence of normal human oral keratinocytes is associated with the loss of telomerase activity witout shortening of telomeres. Cell Growth Differ 9:85–95

Kaouël LBC, Billard C, Macieira-Coelho A (1978) Growth characteristics in vitro of hybrids between normal and transformed cell lines. Int J Cancer 21:338–347

Kempermann G, Kuhn HG, Gage FH (1997) More hipoccampal neurones in adult mice living in an enriched environment. Nature 386:493–495

Kipling D, Cooke HJ (1990) Hypervariable ultra-long telomeres in mice. Nature 347:400–402

Klein B, Pastink A, Odijk H, Westerveld A, van der Erb AJ (1990) Transformation and immortalization of diploid xeroderma pigmentosum fibroblasts. Exp Cell Res 191:256–262

Knees-Matzen S, Roser M, Reimers U, Ehlert U, Weichenthal M, Breitbar EW, Rüdiger HW (1991) Increased UV-induced sister-chromatid exchange in cultured fibroblasts of first-degree relatives of melanoma patients. Cancer Genet Cytogenet 53:265–170

Kopelovich L (1982) Are all normal diploid human cell strains alike? Relevance to carcinogenic mechanisms in vitro. Expl Cell Biol 50:266–270

Koziol C, Borojevic R, Steffen R, Müller WEG (1998) Sponges (Porifera) model systems to study the shift from immortal to senescent somatic cells: the telomerase activity in somatic cells. Mech Ageing Dev 100:107–120

Kraemer PM, Ray FA, Brothman AR, Bartholdi MF, Cram LS (1986) Spontaneous immortalization rate of cultured chinese hamster cells. J Natl Cancer Inst 76:703–709

Kratochwill K (1969) Organ specificity in mesenchymal induction demonstrated in the embryonic development of the mammary gland of the mouse. Dev Biol 20:46–58

Lajtha LG, Schofield R (1971) Regulation of stem cell renewal and differentiation: possible significance in aging. Adv Gerontol Res 3:131–146

Land H, Parada LF, Weinberg RA (1983) Tumorigenic conversion of primary embryo fibroblasts requires at least two cooperating oncogenes. Nature 304:596–598

Littlejohn MR, Camakaris J, Woodcock DM (1995) Cross contamination of the genomes in human/hamster cell hybrids by multiple short recombination events. Somatic Cell Mol Genet 21:385–398

Loo DT, Fuquay JI, Rawson CL, Barnes DW (1987) Extended culture of mouse embryo cells without senescence: inhibiton by serum. Science 236:200–203

Loo DT, Sakai Y, Rowson CL, Barnes DW (1991) Serial passage of human astrocytes in serum-free, hormone supplemented medium. J Neurosci Res 28:101–109

Macieira-Coelho A (1966) Action of cortisone on human fibroblasts in vitro. Experientia 22:390–391

Macieira-Coelho A (1967a) Relationship between DNA synthesis and cell density in normal and virus-transformed cells. Int J Cancer 2:297–203

Macieira-Coelho A (1967b) Dissociation between inhibition of movement and inhibition of division in RSV transformed human fibroblasts. Exp Cell Res 47:193–200

Macieira-Coelho A (1990) Cancer and aging at the cellular level. In: Macieira-Coelho A, Nordenskjöld B (eds) Cancer and aging. CRC Press, Boca Raton, Florida, pp 11–37

Macieira-Coelho A (1991) Chromatin reorganization during senescence of proliferating cells. Mutat Res 256:81–104

Macieira-Coelho A (1994) Genome reorganization through cell division. Implications for aging of the organism and cancer development. Ann N Y Acad Sci 719:108–128

Macieira-Coelho A (1995a) The last mitoses of the human fibroblast proliferative life span, physiopathologic implications. Mech Ageing Dev 82:91–104

Macieira-Coelho A (1995b) Reorganization of the genome during aging of proliferative cell compartments. In: Macieira-Coelho A (ed) Molecular basis of aging. CRC Press, Boca Raton, Florida, pp 21–70

Macieira-Coelho A, Azzarone B (1988) The transition from primary culture to spontaneous immortalization in mouse fibroblast populations. Anticancer Res 8:669–676

Macieira-Coelho A, Diatloff C (1976) Doubling potential of fibroblasts from different species after ionizing radiaiton. Nature 261:586

Macieira-Coelho A, Pontén P (1967) Induction of the division cycle in resting stage human fibroblasts after RSV infection. Biochem Biophys Res Commun 29:316–320

Macieira-Coelho A, Pontén J, Philipson L (1966) The division cycle and RNA synthesis in human diploid cells at different passage levels in vitro. Exp Cell Res 42:673–684

Macieira-Coelho A, Hiu IJ, Garcia-Giralt E (1969) Stimulation of DNA synthesis in resting stage human fibroblasts after infection with Rous sarcoma virus. Nature 222:1172–1173

Macieira-Coelho A, Diatloff C, Billardon C, Bourgeois CA (1977) Effect of low dose rate ionizing radiation on the division potential of cells in vitro. III Human lung fibroblasts. Exp Cell Res 104:215–221

Macieira-Coelho A, Bengtsson A, Van der Ploeg M (1982) Distribution of DNA between sister cells during serial subcultivation of human fibroblasts. Histochemistry 75:11–21

Martinez AO, Norwood TH, Prothero JW, Martin GM (1978) Evidence for clonal attenuation of growth potential in HeLa cells. In Vitro 14:996–971

Masuda N, Fukuoka M, Matsui K, Kudoh S, Negoro S, Takifuji N, Fujisue M, Morino H, Nakagawa K, Nishioka M, Takada M (1991) Establishment of tumor cell lines as an independent prognostic factor for survival time in patients with small-cell lung cancer. J Natl Cancer Inst 83:1743–1747

Matsumura T, Miyashita S, Ohno T (1979) Conversion of proliferation and production of the colony stimulating factor during serial passage of mouse fibroblasts in culture. Cell Struct Funct 4:267–274

McBain JA, Weese JL, Meisner LF (1984) Establishment and characterization of human colorectal cancer cell lines. Cancer Res 44:5813–5821

McCormick JJ, Maher VM (1988) Towards an understanding of the malignant transformation of diploid human fibroblasts. Mutat Res 199:273–291

Meltzer PS, Guan XY, Trent JM (1993) Telomere capture stabilizes chromosome breakage. Nat Genet 4:252–255

Mikol YB, Lipkin M (1984) Methionine dependence in skin fibroblasts of humans affected with familial colon cancer or Gardner's syndrome. J Natl Cancer Inst 72:19–24

Milo GE, Casto BC (1986) Conditions for transformation of human fibroblast cells: an overview. Cancer Lett 31:1–15

Milo GE, DiPaolo JA (1978) Neoplastic transformation of human diploid cells in vitro after chemical carcinogen treatment. Nature 275:130–132

Miranda AF, Duigon GJ, Hernandez E, Fisher PB (1988) Characterization of mutant human fibroblast cultures transformed with simian virus 40. J Cell Sci 89:481–487

Mondal S, Heidelberger C (1977) Transformation of C3H/10T1/2C18 mouse embryo fibroblasts by ultraviolet radiation and phorbol ester. Nature 260:710–712

Morgan TL, Yang D, Fry DG, Hurlin PJ, Kohler SK, Maher VM, McCormick JM (1991) Characteristics of an infinite life span diploid human fibroblast cell strain and a near-diploid strain arising from a clone of cells expressing a transfected v-myc oncogene. Exp Cell Res 197:125–136

Morrison SJ, Wandycz AM, Akashi K, Globerson A, Weissman L (1996) The aging of hematopoietic stem cells. Nat Med 2:1011–1016

Moy EL, Duncan EL, Hukku B, Reddel RR (1997) Reassessment of immortalization complementation group D. Exp Gerontol 32:663–670

Muggleton AL, Danielli JF (1968) Inheritance of the life–spanning phenomenon in Amoeba proteus. Exp Cell Res 49:116–120

Muggleton-Harris AL, DeSimone L (1980) Replicative potentials of various fusion products between WI-38 and SV40 transformed WI-38 cells and their components. Somatic Cell Genet 6:689–695

Mukherji B, MacAlister TJ, Guha A, Gillies CG, Jeffers DC, Slocum SK (1984) Spontaneous in vitro transformation of human fibroblasts. J Natl Cancer Inst 73:583–587

Mullaart EP, van der Lohman PHM, Vijg J (1988) Differences in pyrimidine dimer removal between rat skin cells in vitro and in vivo. J Invest Dermatol 90:346–352

Müller WA (1996) Pattern formation in the immortal Hydra. Trends Genet 12:91–96

Olovnikov AM (1973) A theory of marginotomy. The incomplete copying of template margin in enzymic synthesis of polynucleotides and biological implication of the phenomena. J Theor Biol 41:181–190

Overturf K, Al-Dhalimy M, Ou C-N, Finegold M, Grompe M (1997) Serial transplantation reveals the stem–cell-like regenerative potential of adult mouse hepatocytes. Am J Pathol 151:1273–1280

Panem S, Prochownik EV, Reale FR, Kirsten WH (1975) Isolation of type C virons from a normal human fibroblast strain. Science 189:297–299

Parshad R, Sanford KK, Jones GM (1983) Chromatid damage after G2 phase X-irradiation of cells from cancer-prone individuals implicates deficiency in DNA repair. Proc Natl Acad Sci USA 80:5612–5616

Paules RS, Levedakou EN, Wilson SJ, Innes CL, Rhodes N, Tlsty TD, Galloway DA, Donehower LA, Teinsky MA, Kaufmann WK (1995) Defective G2 checkpoint function in cells from individuals with familial cancer syndromes. Cancer Res 55:1763–1773

Pereira-Smith O, Smith JR (1983) Evidence for the recessive nature of cellular immortality. Science 221:964–966

Pereira-Smith O, Smith JR (1988) Genetic analysis of indefinite division in human cells: identification of four complementation groups. Proc Natl Acad Sci USA 85:6042–6046

Pereira-Smith O, Robetorye S, Ning Y, Orson FM (1990) Hybrids from fusion of normal human T lymphocytes with immortal cells exhibit limited life span. J Cell Phys 144:546–549

Phillips RL, Kaepler SM, Olhoft P (1994) Genetic instability of plant tissue cultures: breakdown of normal controls. Proc Natl Acad Sci USA 91:5222–5226

Pontén J (1971) Spontaneous and virus induced transformation in cell culture. Springer, Berlin Heidelberg, New York

Potten CS, Loeffler M (1990) Stem cells: attributes, cycles, spirals, pitfalls and uncertainties. Lessons for and from the crypt. Development 110:1001–1020

Puck TT, Cieciura SK, Robinson A (1958) Genetics of somatic mammalian cells. III Long-term cultivation of euploid cells from human and animal subjects. J Exp Med 108:945–956

Pye D, MacGregor A, Stanley JF (1977) Marsupial cells in long-term culture. In Vitro 13:232–236

Reynolds CP, Frenkel EP, Smith RG (1980) Growth characteristics of neuroblastoma in vitro correlate with patient survival. Trans Assoc Am Phys 93:203–211

Rogalla P, Rohen C, Henning Y, Deichert U, Bonk U, Bullerdiek J (1995) Telomere repeat fragment sizes do not limit the growth potential of uterine leiomyomas. Biochem Biophys Res Commun 211:175–182

Rothstein R, Gangloff S (1995) Hyper-recombination and Bloom's syndrome: microbes again provide clues about cancer. Genome Res 5:421–426

Rubin H (1997) Cell aging in vivo and in vitro. Mech Ageing Dev 98:1–35

Ryan PA, Maher VM, McCormick JJ (1994) Failure of infinite life span human cells from different immortality complementation groups to yield finite life span hybrids. J Cell Phys 159:151–160

Sager R, Tanaka K, Lan CC, Ebina Y, Anisowicz A (1983) Resistance of human cells to tumorigenesis induced by cloned transforming genes. Proc Natl Acad Sci USA 80:7601–7605

Salk D, Bryant E, Au K, Hoehn H, Martin G (1981) Systematic growth studies, cocultivation and cell hybridization studies of Werner syndrome cultured skin fibroblasts. Hum Genet 58:310–321

Schmitt H, Blin N, Zankl H, Schertan H (1994) Telomere length variation in normal and malignant human tissues. Genes Chromosomes Cancer 11:171–177

Schneider EL, Gilman B (1979) Sister chromatid exchanges and aging. III. The effect of donor age on mutagen-induced sister chromatid exchanges in human diploid fibroblasts. Hum Genet 46:57–62

Schor SL, Schor AM, Rushton G (1988) Fibroblasts from cancer patients display a mixture of both foetal and adult-like phenotypic characteristics. J Cell Sci 90:401–407

Schor SL, Schor AM, Grey AM, Chen J, Rushton G, Grant ME, Ellis I (1989) Mechanism of action of the migration stimulating factor produced by foetal and cancer patient fibroblasts: effect on hyaluronic acid synthesis. In Vitro Cell Dev Biol 25:737–746

Schultz VP, Zakian VA, Ogburn CE, McKay J, Jarzebowicz AA, Edland SD, Martin GM (1996) Accelerated loss of telomeric repeats may not explain accelerated replicative decline of Werner syndrome cells. Hum Genet 97:750–754

Schwartz AG, Moore CJ (1977) Inverse correlation between species life span and capacity of cultured fibroblasts to bind 7,12-dimethyl benzaanthracene to DNA. Exp Cell Res 109:448–452

Shay JW, Wright WE, Werbin H (1991) Defining the molecular mechanisms of human cell immortalization. Biochim Biophys Acta 1072:1–7

Shima A, Nikaido O, Shinohara S, Egami N (1980) Continued in vitro growth of fibroblast-like cells (RBCF-1) derived from the caudal fin of the fish, Carassius auratus. Exp Gerontol 15:305–314

Sigel MM, Beasley AR (1973) Marine teleost fish tissues. In: Kruse PF Jr, Patterson MK (eds) Tissue culture, methods and applications. Academic Press, New York, pp 123–128

Smith HS, Wolman SR, Dairkee SH, Hancock MC, Lippman M, Leff A, Hackett AJ (1987) Immortalization in culture: occurrence at a late stage in the progression of breast cancer. J Natl Cancer Inst 78:611–615

Spandidos DA, Wilkie NM (1984) Malignant transformation of early passage rodent cells by a single mutated human oncogene. Nature 310:469–471

Stanley JF, Pye D, MacGregor A (1975) Comparison of doubling numbers attained by cultured animal cells with life span of the species. Nature 255:158–159

Stehelin D, Guntaka RV, Varmus HE, Bishop JM (1976) Purification of DNA complementary to nucleotide sequences required for neoplastic transformation of fibroblasts by avian sarcoma viruses. J Mol Biol 101:349–356

Stenkvist B (1966) Long-term cultivation of human and bovine fibroblastic cells morphologically transformed in vitro by Rous sarcoma virus. Acta Pathol Microbiol Scand 67:67–82

Stewart FC (1963) The control of growth in plant cells. Sci Am 209:104–112

Suda Y, Suzuki M, Ikawa Y, Aizawa S (1987) Mouse embryonic cells exhibit indefinite proliferative potential. J Cell Phys 133:197–201

Swim HE, Parker RF (1957) Culture characteristics of human fibroblasts propagated serially. Am J Hyg 66:235–243

Todaro GJ, Green H (1963) Quantitative studies of the growth of mouse embryo cells in culture and their development into established cell lines. J Cell Biol 17:299–313

Torrey JG (1967) Morphogenesis in relation to chromosomal constitution in long-term plant tissue cultures. Physiol Plant 20:265–275

Tropene V, Craig CG, Morshead CM, van der Kooy D (1997) Stem cells in the mammalian brain. J Neurosci 15:7850–785

Tsutsui T, Fujino T, Kodama S, Tainsky MA, Boyd J, Barrett JC (1995) Aflatoxin B1-induced immortalization of cultured skin fibroblasts from a patient with Li-Fraumeni syndrome. Carcinogenesis 16:25–34

Tsutsui T, Tanaka Y, Matsudo Y, Hasegawa K, Fujino T, Kodama S, Barrett JC (1997) Extended life span and immortalization of human fibroblasts induced by X–ray irradiation. Mol Carcinog 18:7–18

Tweeddale ME, Lim B, Jamal N (1987) The presence of clonogenic cells in high-grade malignant lymphoma: a prognostic factor. Blood 69:1307–1314

Varma M, Leavitt J (1988) Macromolecular changes accompanying immortalization and tumorigenic conversion in a human fibroblast model system. Mutat Res 199:437–447

Vaziri H, Dragowska W, Allsop RC, Thomas TE, Harley CB, Lansdorp PM (1994) Evidence for a mitotic clock in human hematopoietic stem cells: loss of telomeric DNA with age. Proc Natl Acad Sci USA 91:9857–9860

Vojta PJ, Barrett JC (1995) Genetic analysis of cellular senescence. Biochim Biophys Acta 1242:29–41

Vos O, Dolmans MJAS (1972) Self-renewal of colony forming units (CFU) in serial bone marrow transplantation experiments. Cell Tissue Kinet 5:31–385

Weissmann A (1891) Essays upon heredity and kindred biological problems. Clarendon Press, Oxford

Wilmut I, Schnieke AE, McWhir J, Kind AJ, Campbell HS (1997) Viable offspring derived from fetal and adult mammalian cells. Nature 385:810–813

Xia SJ, Shammas MA, Shmookler Reis RJ (1997) Elevated recombination in immortal human cells is mediated by HsRAD51 recombinase. Mol Cell Biol 17:7151–7158

Yagi T (1982) DNA repair ability of cultured cells derived from mouse embryos in comparison with human cells. Mutat Res 96:89–94

Yang D, Louden C, Reinhold DS, Kohler SK, Maher VM, McCormick JJ (1992) Malignant transformation of human fibroblast cell strain MSU-1.1 by terahydrobenzo[a]pyrene. Proc Natl Acad Sci USA 89:2237–2241

Yao M-C, Choi J, Yokoyama S, Austerberry CF, Yao C-H (1984) DNA eliminaton in **tetrahymena**: a developmental process involving extensive breakage and rejoining of DNA at defined sites. Cell 36:433–441

Clonal Life Cycle of *Paramecium* in the Context of Evolutionally Acquired Mortality

Y. Takagi[1]

1
From Immortality to Mortality

A single bacterium dividing every 20 minutes, if nutrients could be supplied, would divide 144 times in 2 days to produce 2^{144} (10^{43}) bacteria. Their weight would exceed the weight of the earth (6×10^{27} g), even if the weight of a single bacterium is underestimated at 10^{-15} g ($10^{-15} \times 10^{43} = 10^{28}$). Our body, consisting of on the order of 10^{13} cells, all of which have been derived from a single cell (a fertilized egg) would be attained by only 43 cell divisions and would become the size of an elephant after several more cell divisions, if a fertilized egg grew exponentially ($2^{43} = 10^{13}$). These examples of the tremendous power of exponential cell division indicate how important it is for living organisms to regulate cell division.

Until about the middle of this century, the cells of both unicellular and multicellular organisms were thought to have the potential to divide indefinitely. This was disproven by Sonneborn (1954) in *Paramecium* and by Hayflick et al. (1961) in normal human cells cultured in vitro; both kind of cells age with cell division and die after a definite number of cell divisions, called the "Sonneborn limit" for *Paramecium* cells and "Hayflick limit" for human cells.

Procaryotes such as bacteria and archaea, which constituted the sole group of living organisms on the earth during the first two-thirds of evolutionary history until eucaryotes evolved from them, are thought to be potentially immortal; they stop dividing only when food resources become unavailable, whereas eucaryotes stop cell division at the Sonneborn or Hayflick limit even in nutrient-rich conditions. From the evolutionary viewpoint, therefore, the fundamental nature of cells is immortal, and mortality is an evolutionary phenomenon acquired by eucaryotes. Metaphorically speaking, a primitive car not equipped with a brake has evolved to a modern car equipped with a brake and an accelerator whereby the car can stop even when fuel is not exhausted or when it is on a slope.

In addition to the ability to stop cell division, the ability to die (programmed cell death or apoptosis) is also thought to have been acquired by

[1] Department of Biology, Nara Women's University Kita-uoya Nishi, Nara 630–8506, Japan.

Progress in Molecular and Subcellular Biology, Vol. 24
A. Macieira-Coelho
© Springer-Verlag Berlin Heidelberg 1999

eucaryotes, although intracellular processes something like apoptosis have been suggested to occur even in procaryotes (Ameisen 1998). In unicellular eucaryotes, however, apoptosis-like processes have been reported only in some organisms such as *Trypanosoma, Leishmania, Dictyostelium*, and *Tetrahymena* (see review by Ameisen 1998). Needless to say, apoptosis in multicellular organisms is essential for morphogenesis during ontogeny and for homeostasis throughout their lifetimes.

Somatic cells in the animal body can be roughly classified into 2 types, those that have the ability to proliferate, irrespective of whether that ability is utilized or not, and those that lack the ability to proliferate. The adult bodies of vertebrates are composed of these 2 types, those of some invertebrates such as insects, nematodes and rotifers are composed of only non-proliferative cells, and those of some others such as coelenterates and plathelminthes are composed of only proliferative cells. In all of them, the early embryos are composed of proliferative cells. Both phylogenetically and ontogenetically, therefore, the direction of cell division potential is from proliferative to non-proliferative. When mortal cells (e.g., normal human diploid cells) and immortal cells (e.g., cancer cells) are fused, the resulting fused cells become mortal (Bunn et al. 1980; Muggleton-Harris et al. 1980; Pereira-Smith et al. 1983), showing that mortality is dominant over immortality.

Altogether, it may be that eucaryotes have set up regulation systems inhibitory to the immortal mechanisms inherited from procaryotes. I presented in my book written in Japanese (Takagi 1993) some basic ideas arguing that immortality is the "default" for the cell, the *brake system* was invented in eucaryotes, and both the differentiation into germ and soma and the sexual reproduction represented by meiosis and fertilization (regarding recombination as optional) may involve the brake system. I will discuss in this chapter how the cellular lifespan has evolved, and try to apply the implications to our investigations using the unicellular protozoan *Paramecium* as a model organism.

2
From Horizontal to Vertical Genetic Axis

Gene transfer from cell to cell through cell division, i.e., asexual reproduction, is the fundamental process of genetics. The genetic axis of this horizontal inheritance is characterized by uniformity; genes are copied by means of the semiconservative DNA replication mechanism to result in genetically uniform cells forming a clone. Practically all of the procaryotes and some eucaryotic unicellular organisms have only this axis: they repeat horizontal inheritance indefinitely. In adverse environments, they can survive either by a makeshift mechanism (a transient stop of cell division usually followed by development of defensive structures such as spores) or by elimination of a part of the clonal population that failed to adapt to the environment (Zambrano et al. 1993).

Most multicellular organisms have another axis, vertical inheritance: gene transfer from one generation to the next generation. The core process of this axis is meiosis (2 n → n) and fertilization (n → 2 n). It is only when this axis was introduced that clonal termination became possible (the horizontal axis came to be mortal). The vertical axis is characterized by heterogeneity, allowing genes to be recombined or shuffled, although gene exchange or crossing over is not necessarily essential for the process of meiosis and fertilization. As for the diploid eucaryotes, the vertical axis, if repeated, represents phylogeny, while the horizontal axis represents ontogeny.

3
From Small to Large Cells

Eucaryotic cells, which have an average diameter of about 10 μm, are 1,000 times as large in volume as procaryotic cells, which have an average diameter of about 1 μm. The well-known characteristics of eucaryotic cells appear to be closely related to this dramatic increase in their size: genetic material (DNA) should increase by 1,000 times to drive and maintain the 1,000-fold larger cell, and the enlarged DNA should be safely stored, not entangled (DNA packaging using histones to compact DNA into nucleosomes, which are in turn stored in the nucleus); supporting and transportation systems should develop for the 1,000-fold larger cell (microtubules and microfilaments); cell membranes should extend inside to compensate for the lack of cell surface, which would increase only by 100 times (well-developed intracellular membrane systems are used not only for gas exchange but also for transporting channels).

The eucaryotic cell has further enlarged in some lineages, including the unicellular ciliates, which are about 100 μm in diameter and thus 1,000,000 times larger than procaryotes, while in cell lineages that have evolved into the multicellular organisms, the individual cell size has remained unenlarged.

Supposing that an ancestral procaryote with a single-copy genome increased the genetic material concomitantly with the cell volume, the increased DNA would require more time for its replication and become susceptible to more chance of mutations. Most mutations in the single-copy genome would result in the death of their possessors, unless they were protected by safety devices. Besides the nucleosome structure stored in the nucleus, the following two may be of special importance as safety devices;

(1) differentiation of the cells into the treasured "germ", which inherited the immortal nature of procaryotes and the "soma", which can utilize costly energy for work other than cell division, and

(2) possession of genetic material in duplicate, i.e., genomic diploidization. These will be discussed in the following section in more detail.

4
A Possible Scenario of the Evolution of Cellular Lifespan

The scenario will be summarized as a proposal that the *safety devices*, i.e., "differentiation into germ and soma" and "diploidization", are coupled to the interrelated *brake systems* to control replication in soma, to induce apoptotic degradation of soma, and to initiate meiosis in germ.

4.1
Differentiation into Germ and Soma

The original immortal cells are alternatively called "germ" cells. When "soma" cells were produced for the first time among the germ cell population, it appears that they were destined to die without cell division. Therefore, the terms germ and soma are, at this stage, identical to proliferative and non-proliferative, respectively, and the term germ has nothing to do with sexual reproduction. Differentiation into germ and soma is possible only in a multicellular (or multinuclear) state. Since procaryotes remain potentially immortal, although they can assume a multicellular state (Kolter et al. 1998), this differentiation may have occurred after eucaryotes evolved. Since this differentiation is seen in haploid eucaryotes if they form a colony or aggregate to a multicellular form, this differentiation may have occurred before diploidization evolved in eucaryotes. The differentiation into proliferative germ and non-proliferative soma is observed in some present-day haploid eucaryotes: for example, colony-forming phytoflagellates such as *Eudorina*, *Pleodrina* and *Volvox* differentiate proliferative and non-proliferative cells in a colony; amoebas such as *Dictyostelium* have a multicellular stage in their life cycle and differentiate proliferative cells (spore) and non-proliferative cells (stalk) (see Grell 1973).

Later in the evolutionary process, probably after diploid cells evolved, the germ came to be associated with sexual reproduction, thus introducing the vertical genetic axis, and the soma came to have a regulated proliferative ability. Thus, both germ and soma are proliferative, but the proliferative ability of the soma is limited. This is what we call cellular lifespan. In this scenario, the soma, which had been destined to die in haploid eucaryotes, gained the ability to proliferate in diploid eucaryotes. Because genetic traits once lost during evolution are rarely regained, it may be that the proliferative ability was not lost in haploid eucaryotes, but only suppressed.

In almost all ciliates, in which differentiation into the germ (diploid micronucleus) and soma (polygenic macronucleus) occurs at the nuclear level, and in which both the germ and soma are proliferative, the germ is produced from the germ and the soma from the soma during asexual reproduction. During sexual reproduction, both the germ and soma are produced from the germ passing through meiosis, mitosis, fertilization and mitosis (the old soma is destroyed).

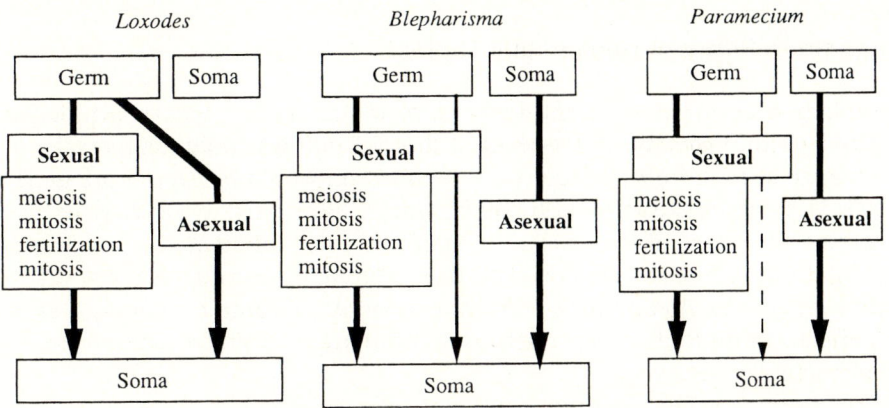

Fig. 1. Formation of soma (macronucleus) from germ (micronucleus) or from soma in 3 ciliates. In *Loxodes*, soma is always formed from germ during sexual and asexual reproductions. In *Blepharisma* and *Paramecium*, soma is formed from soma during asexual reproduction, and soma is formed from germ during sexual reproduction passing through a series of meiosis, mitosis, fertilization and mitosis. In addition to this main pathway during sexual reproduction (*thick line*), soma can be formed directly from germ in *Blepharisma* (*thin line*), while in *Paramecium*, formation of soma can be induced only by artificial means (*broken line*)

In the karyorelictids (e.g., *Loxodes*), which are thought to belong to a primitive group of ciliates and in which the macronucleus has a diploid G_2 amount of DNA (Raikov 1982) and is still non-proliferative, the soma is formed from the germ not only during sexual reproduction but also during asexual reproduction (Fig. 1). The old soma is not destroyed, either during sexual reproduction or during asexual reproduction (see Orias 1986). During sexual reproduction in the heterotrichid ciliate *Blepharisma*, the soma is produced from the germ either by passing through a series of meiosis, mitosis, fertilization and mitosis or by direct differentiation of the germ (Miyake et al. 1991). Even in *Paramecium*, the soma can be produced directly from the germ if the germ is artificially placed in exconjugant cytoplasm in which the macronucleus is differentiating from the fertilized micronucleus (Mikami et al. 1983b). These features suggest that for the production of the proliferative (to a limited extent) soma from the germ, what is essential is not an experience of "undergoing meiosis and fertilization" but one of "being placed in a spatio-temporally specific field" which is created through meiosis and fertilization. At least in ciliates, the mechanism for initiating meiosis in the germ seems to be coupled to the mechanism for destroying the soma (apoptotic degradation of the old macronucleus).

4.2
Alternation Between Haploid and Diploid

In enlarged eucaryotes, harmful effects of mutations are mostly avoided if diploidization takes place. However, if diploidization is only for the sake of avoidance of mutational effects, polyploidization would have even more beneficial. Diploidy instead of polyploidy would have been preferred for the selective advantage of the ease of reverting to the haploid state.

Mutations are resources for evolution, although harmful in most cases. Some mutations, if combined with others, would become harmless or even beneficial. In diploid cells, numerous combinations of mutations are stored in one genome set.

However, recessive mutations are not expressed unless their possessors become haploid or are recessive homozygotes. The most primitive form of meiosis might have been one-step haploidization through "genomic segregation", as depicted in Fig. 2-Ia. The transition of n → 2 n through genomic duplication (A) and that of 2 n → n through genomic segregation (B) are the regular process of "mitosis" in the haploid cell. If the same process involving genomic duplication and genomic segregation is performed in the diploid cell, it is nothing else than the normal process of "mitosis" in the diploid cell (Fig. 2-II). What is unique in "meiosis" is the "pairing and segregation of the homologous chromosomes" (Fig. 2-Ib, D, E). The "chromosomal recombination" may be optional. The two-step genomic reduction of "meiosis" in modern multicellular organisms (Fig. 2-III) can be viewed as the process in which the chromosomal pairing (D) and segregation (E) are intercalated into the processes of the genomic duplication (A) and segregation (B).

4.3
When Do Diploid Cells Become Haploid?

Alternation between haploid and diploid constitutes the regular process of mitosis for haploid cells (Fig. 2-I), but would later become a well-controlled process in diploid cells. The controls involve when and how haploidization occurs. In both paths of diploidization, i.e., genomic duplication in and homotypic cell fusion of haploid cells, the two sets of genomes would remain identical unless mutations were stored in one of them. In this situation, chromosomal exchange or crossing over does not produce genetic variations, so that there is no selective advantage to employing recombination at the DNA level so long as alternations between haploid and diploid are occurring as frequently as elimination of the accumulation of mutations.

In order to make full use of the strategy, "store sufficient mutations and then test their availability", the timing of haploidization should be controlled. The duration of the diploid state until the time of haploidization, i.e., the duration for sexual maturity, would have been subjected to natural selection. It may depend on the rate of mutation accumulation, the rate of repair

Fig. 2. Two possible patterns of transition from haploid to diploid and from diploid to haploid (**Ia** and **Ib**), and schematic view of mitosis (**II**) and meiosis (**III**) in the diploid cell. *Open* and *shaded square bars* show homologous chromosomes or a genome. A: genomic duplication, B: genomic segregation, C: cell fusion, D: pairing of and recombination (X) between the homologous chromosomes, E: segregation of the homologous chromosomes. Note that mitosis in diploid cells is the process in which D and E are intercalated between A and B; **Ia** is the common process of mitosis in a haploid cell, inferring that mitosis potentially includes a meiotic nature; chromosomal recombination is not necessarily essential for meiosis

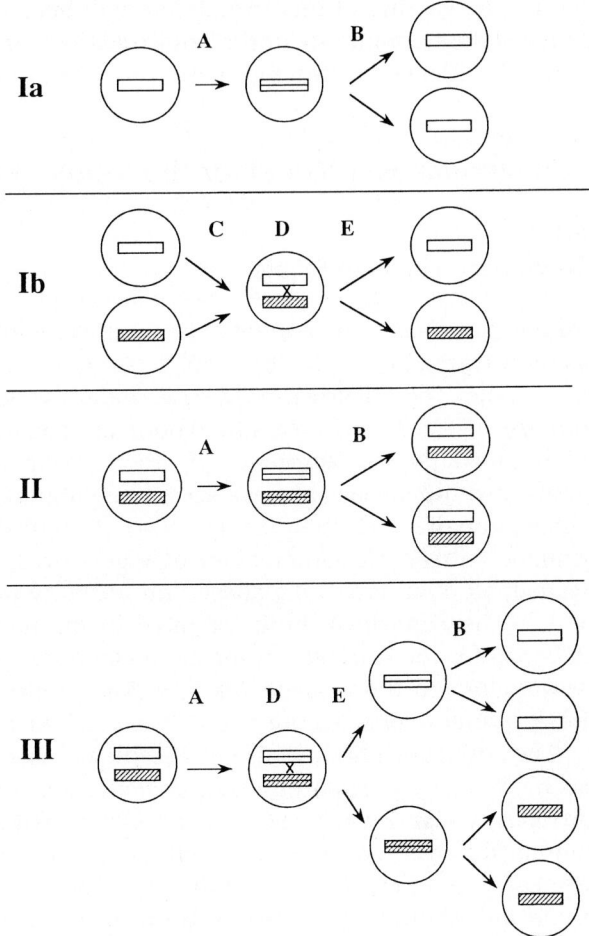

efficiency, the amount of genome size, the breeding strategy such as inbreeder or outbreeder and so on.

In the diploid cells which differentiate into germ and soma, haploidization takes place in the germ line. For the diploid germ cells, to become haploid means that the continuity of their diploidy ends. This contradicts the immortality of the horizontal genetic axis in the germ line. On the other hand, the diploid soma cells do not have to stop dividing to become haploid, so that they can theoretically be immortal! This contradiction should be resolved in some way. My speculation is that the mechanism for the interruption of cell division in diploid germ cells to promote meiosis was utilized as the brake system to limit the division potential in diploid soma cells. This indicates that the mechanism of limiting the cellular lifespan in soma might have developed in close association with the mechanism of sexual reproduction in germ; in other words, the brake system to suppress cell division in soma might have coupled to the accelerator system to initiate meiosis in germ. The validity of this speculation will become clear if conserved genes that function commonly in the both processes are identified. Some circumstantial evidence for this speculation will be discussed in the final section.

5
Paramecium as a Model for the Study of Cellular Lifespan

5.1
Sonneborn Limits in Ciliates

Paramecium was once thought to be immortal (able to multiply indefinitely), as represented by the "Methuselah" strain of one of the species of the *P. aurelia* complex (designated now as *P. biaurelia*), which had been cultured without loss of vigor for 33 years by Woodruff of Yale University. The possibility of immortality was disproven by the finding of periodic occurrence of autogamy, the nuclear process accompanying meiosis and fertilization in a single, unpaired cell (Sonneborn 1954). Also in *P. caudatum*, in which continuous cultivation without loss of vigor for 22 years had been reported (Galadjieff et al. 1933), the alleged immortality was disproven (Takagi et al. 1980). The Sonneborn limits counted in the number of cell divisions are: 200–350 for the *P. aurelia* complex, including *P. primaurelia, P. biaurelia, P. tetraurelia* and *P. octaurelia*, 500–750 for *P. caudatum*, and longer than that for *P. multimicronucleatum* and *P. bursaria* (for review, see Takagi 1988).

The Sonneborn limit is also applicable to ciliates other than *Paramecium*, such as *Tetrahymena, Euplotes, Stylonychia, Oxytricha, Spathidium* and *Tokophrya* (for review, see Smith-Sonneborn 1985). It is highly variable among different genera, among different species of the same genus, and even among different strains of the same species: about 200 in fresh-water *E. woodruffi*, about 400 in marine strains of the same species, about 700 in *E. minuta*, 1,000–1,500 in *E. crassus* and *E. patella*, 300–1,000 in *Stylonychia* and 500–1,500 in *Tetrahymena*.

The sole exception is the amicronucleate strain GL of *Tetrahymena pyriformis*, which is genetically dead because of the lack of a germinal micronucleus and yet immortal in that it can multiply indefinitely (Nanney 1974).

5.2
Time-Measuring Mechanisms in Ciliates

5.2.1
Spiral Cell Cycle

During vegetative reproduction, the cells of ciliates that have successfully accomplished sexual reproduction (conjugation or autogamy) undergo unidirectional changes of immaturity, maturity and senescence terminating with clonal death. *Paramecium* cells use the number of cell divisions instead of days to count the timing of sexual maturation (Kroll et al. 1968; Takagi 1970; Miwa et al. 1970; Ishikawa et al. 1998). This is also true for the timing of clonal death (Smith-Sonneborn et al. 1976; Takagi et al. 1980; Takagi et al. 1987a). More precisely, the number of DNA replications rather than the number of cell divisions was shown to be the counter for the onset of sexual maturity (Mikami et al. 1983a), although the two counters basically accord under usual laboratory conditions where one round of the cell cycle is accompanied by one round of DNA replication.

One round of the cell cycle includes, as a rule, one round of DNA synthesis both in the micronucleus and in the macronucleus. The cell cycle usually differs between the two kinds of nuclei; for example, in *Paramecium*, the G_1, S, G_2 and M phases are distinguishable in the micronucleus, while there are only G_1, S and D (division; termed so because of amitosis) phases in the macronucleus, and the lengths of the respective phases in the two nuclei do not correspond to each other (Berger 1988). Further, successive macronuclear cell cycles differ in terms of the lengths of these phases, and thus they should be viewed as spiral cycles instead of closed circles.

Yamamoto et al. (1997) showed in *P. tetraurelia* that UV sensitivity decreased as cells proceeded through the cell cycle, and thus that G_1 cells, after one round of the cell cycle, returned to the most sensitive state. However, the UV sensitivity in G_1 cells did not return to the original level but increased with increasing rounds of cell cycles, confirming the earlier report by Smith-Sonneborn (1971). The time arrow as shown in the change of UV sensitivity was not so simple as expected, however. The increase in UV sensitivity with the advance of clonal age was almost linear only after the clonal age of 80 cell divisions, but clones younger than that age showed sporadic hypersensitivity to UV.

5.2.2
Time Record

Irreversible temporal changes must be recorded somewhere in the cell. The telomere shortening hypothesis (Olovnikov 1973) has attracted many scientists who are interested in the molecular clock that triggers senescence (Harley et al. 1990; Harley 1991; Greider 1996). At least in human diploid fibroblasts cultured *in vitro*, a causal relationship was established between telomere shortening and cellular senescence (Bodnar et al. 1998). In *Paramecium*, however, no telomere shortening was detected during clonal aging; instead, the size of DNA became smaller with clonal age (Gilley et al. 1994). Although the change in the molecular size of DNA does not infer a change in the total DNA content, the macronuclear DNA content in *Paramecium* has been reported to decrease with clonal age (Schwartz et al. 1973, 1975; Klass et al. 1976; Takagi et al. 1982). The pattern of decreasing DNA content is, however, not consistent either among different species or with the pattern of decreasing molecular size of the DNA.

Another candidate for the time-recording place is the cell cortex. Once formed, the cortex of the *P. aurelia* complex is persistent and nonrenewable during asexual reproduction, while the cortex of *T. pyriformis* grows and is replaceable when damaged (Sonneborn 1974b). Sonneborn (1974b) related this cortical difference to mortality in the *P. aurelia* complex and immortality in *T. pyriformis*, but I speculate differently. The existence of immortal species in ciliates may provide supporting evidence for my speculation mentioned above: the mechanism that initiates meiosis is related to the mechanism that sets the cellular lifespan. In *T. pyriformis*, in which the germinal micronucleus is lost and the somatic macronucleus divides indefinitely, the mechanism that sets the clonal lifespan may be missing in the macronucleus and thus the coupling mechanism that initiates meiosis is also missing, so that the micronucleus may have been lost as a result of selective pressure against the burden of its maintenance. I suppose, as will be discussed in the final section, that even *Paramecium* cells would become immortal if the mechanism of setting the lifespan could be canceled. This will never be accomplished by elimination of the micronucleus, which, instead, usually results in weak, sterile cells. It should be noted that the macronucleus possesses the mechanism controlling the clonal lifespan and meiosis irrespective of the presence or absence of the micronucleus.

5.2.3
Changing Disorder to Order

The irreversible temporal changes during the clonal life cycle that occur in a predictable way under uniform conditions suggest the existence of some intrinsic time-measuring mechanisms, making them comparable to developmental changes in vertebrates. However, random events such as accumulation of defects in the nuclei and in the cytoplasm or random distribution of macronuclear DNA in an amitotically dividing macronucleus might also be used for long-term temporal regulation (Nanney 1974). Senescence in *Paramecium* is accompanied by increasing defects in the macronuclei, micronuclei and cytoplasm (Sonneborn 1974b; Smith-Sonneborn 1981, 1985; Takagi 1988). Although the ciliates' phenotypes are largely controlled by the macronucleus, there is some experimental evidence suggesting that the primary senescent changes are not in the nuclei but in the cytoplasm (Sonneborn et al. 1960a, 1960b, 1960c; Karino et al. 1981, 1984). In most cases, however, the cytoplasm is characterized by the products of macronuclear genes, except for the mitochondria and the cortex. The microinjection of cytoplasm between young and old cells did not modify the clonal lifespan of the recipient cells, indicating that the cytoplasm contributes little, if any, to the clonal aging (Aufderheide 1984).

What has been most extensively discussed as a time-measuring mechanism is the dilution hypothesis, which supposes that some inhibitors or activators stored in a starting cell might be diluted by cell division to a threshold level. Since the simple dilution model must postulate too much of a substance to account for exponential dilution, a modified form of the differential production model has been proposed (Miwa et al. 1970; Nanney 1974, 1980); this postulates the synthesis of inhibitors or activators during vegetative growth but at a rate somewhat lower than the rate of dilution. These considerations led to the finding of a novel protein, "immaturin", that inhibits expression of the mating reaction and thus keeps the cells in an immature state (Miwa et al. 1975; Haga et al. 1981; Miwa 1984).

As exemplified in telomere shortening, a parental double strand of DNA does not yield strictly identical progeny DNA: the daughter strand that lacks the 5' primer region is shorter than the parental strand; the lagging strand may store more mutations than the leading strand because of the difference of biochemical complexity involved in producing each strand (Alberts et al. 1994). This means that the cells of a single clone are heterogeneous in, for example, how many mutations are stored in their DNA. Therefore, genetically different cell lineages may be produced intraclonally. Since the macronucleus of ciliates is polygenic, a mutation-based clock may be feasible especially for the senescence-related temporal phenomena, if there is a threshold level of mutations beyond which a phenotype can not be maintained.

The ciliate macronucleus divides amitotically, lacking a segregation-based mechanism to ensure that each daughter cell receives an identical set of chromosomes. A random partitioning of macronuclear DNA at every cell division (and yet genetically regulated; see Berger 1988) may produce heterogeneous cells with different amounts of macronuclear DNA. The random partitioning of macronuclear DNA may be used as a cellular clock, if a threshold level of DNA content is needed to maintain a phenotype. In *Paramecium,* a gross deviation from equality of the macronuclear DNA in daughter cells has been suggested to be associated with the initiation of senescence (Takagi et al. 1982).

In *Tetrahymena,* the macronucleus is composed of 45 units of the haploid genome set, which is randomly distributed to daughter cells, and thus, progeny heterozygous for a dominant and recessive allelic pair expresses the dominant phenotype at first, but yields subclones stably expressing the recessive phenotype. This phenomenon, called *phenotypic assortment,* has been used to create strains which express a particular phenotype (Sonneborn 1974a; Nanney 1980; Orias 1998).

The ciliate macronucleus is, however, greatly diversified not only in shape, size and number but also in chromatin structure, genic composition, and replication and partitioning of DNA, making it difficult to apply information obtained in one ciliate to another (Karrer 1986; Klobutchr et al. 1986; Freiburg 1988; Raikov 1995).

5.2.4
Modification of the Lifespan

Another approach to discovering the time-measuring mechanism is to study by what means the clonal lifespan is modified. Several factors have been found so far. As expected, those that are known to induce malfunctions, such as UV, reduce the lifespan (Smith-Sonneborn 1971), and those that are known to counteract malfunctions, such as vitamin E, extend the lifespan (Thomas et al. 1988). Of special interest is the life-extending effect of visible light irradiation following UV irradiation (UV+PR) (Smith-Sonneborn 1979). Although the mechanism by which this treatment can extend the clonal lifespan remains unknown, Smith-Sonneborn speculated that some cryptic mechanism for repairing DNA damage may be involved.

All the factors known to extend the lifespan are, however, effective for extending the average value, not for extending the species-specific maximum lifespan. If the lifespan is programmed by genes, alterations of these genes may cause shortening, or theoretically even elongation, of the maximum lifespan. As discussed previously, the "default" for the cells is "immortal", and the "brake system" has been "invented" through evolution. If we change the set-point of the brake, the cells may modify the life cycle stage, and if we break or take off the brake, the cells may be immortalized. We have, in fact, the example of the GL strain of *Tetrahymena pyriformis* for immortality. In order to test whether the lifespan is programmed by genes, I have isolated mutants with modified clonal lifespans. One of the mutants with novel life cycle features will be described below.

5.3
Association of the Clonal Lifespan with Other Life Cycle Features

5.3.1
Clonal Lifespan Versus the Length of Sexual Immaturity

The clonal life cycle of ciliates may not be under one consistent kind of temporal control throughout the life, but under two distinct kinds of temporal control. The earlier part of the life cycle, until sexual maturity (development), may be an intrinsic process under precise temporal regulation, while the later part, until death (senescence), may be an extrinsically controlled process that is subject to random events. I think, as mentioned previously, that the mechanism of limiting the ability of the cells to divide was originally the same as the mechanism of committing the cells to initiate meiosis. The two processes, i.e., the initiation of meiosis and the restriction of cell division (cellular lifespan), may have been temporally separated by some regulatory mechanism. The initiation of meiosis to introduce the vertical genetic axis is of vital importance for the cell, so that the regulation systems involved in this process may be under the strict surveillance of natural selection. On the other hand, the regulatory mechanism for the restriction of cell division (cellular lifespan) may become free from the strict surveillance of natural selection, so that lifespan may become a highly variable phenotype. However, the evolutionary track that once connected the two processes may have remained, leading to a degree of correlation between the length of development and the lifespan.

It appears meaningful to refer to a loose coupling between the length of lifespan and the length of development (species that have longer lifespans tend to have longer periods of sexual immaturity) in such remote groups of animals as mammals (Cutler 1978) and ciliates (Smith-Sonneborn 1981). These observations suggest that the mechanism used for measuring the sexual maturation is also used for measuring the clonal lifespan.

In the *jumyo* mutant of *P. tetraurelia,* which has an extremely short clonal lifespan (see the last section for detail), the length of autogamy immaturity is also short (Takagi et al. 1987b). In wild type cells, the length of sexual immaturity (the number of cell divisions until conjugation becomes possible), the length of autogamy immaturity (the number of cell divisions until autogamy becomes inducible by natural starvation), and the length of clonal lifespan are all shortened by starting the new generation from a parent of more advanced clonal age (Siegel 1961; Iizima et al. 1997; Smith-Sonneborn et al. 1974). These carry-over effects of the parental age on the next generation may be related to the efficiency of a clock rewinding mechanism that may work during sexual reproduction.

5.3.2
Clock Rewinding During Sexual Reproduction

The questions of why and how cells are rejuvenated during conjugation and autogamy remain unanswered. The generally accepted idea is that the mutations accumulated in the macronucleus are canceled by degrading it and constructing a new macronucleus from the germ micronucleus. Mutations are also accumulated in the micronucleus, but deleterious mutations in the micronucleus may be eliminated through recombination at meiosis, and the intact genome will be restored in the differentiating macronucleus. In this explanation, accumulation of mutations is the clock and their elimination through recombination is the clock rewinding.

In *P. tetraurelia,* cells of the age of around 20 cell divisions can be reset to time zero through autogamy. If they repeat autogamy every 25 cell divisions, they can live indefinitely. According to Sonneborn (1974b), the frequency of death at autogamy was only 0.7 % on average in 50 successive autogamies induced every 25 cell divisions of parent clones. Note that autogamy is the process that occurs in a single cell and consists of meiosis, mitosis, fertilization and mitosis of the micronucleus to result in homozygosity in all genetic loci. If mutations do not occur in the micronucleus during 2 successive autogamies, the parent and the autogamy-progeny are identical in both their genetic and cytoplasmic compositions, and yet different in their clonal ages. The clock rewinding in this case has nothing to do with mutation canceling.

Then, what kind of mechanism for clock rewinding is possible? My idea is that the clock rewinding in the nucleus will be done through interactions with the spatio-temporally specific cytoplasm. Although the mechanism is yet unknown, it should work sometime during nuclear construction during autogamy (or conjugation), including the period encompassing prezygotic and postzygotic micronuclear changes, development of the new macronucleus (anlage), and degradation of the old macronucleus. The cytoplasm as well undergoes dramatic changes spatially and temporally during this period (for review see Fujishima 1988; Mikami 1988). Since what is actually rejuvenated is the macronucleus, which will govern the phenotypes of the progeny, nucleo-cytoplasmic interactions via postzygotic micronuclear changes appear to be pivotal. The nuclear changes during this period are: the fertilized nucleus divides mitotically 2 times to produce 4 products, 2 of which are located in the posterior region of the cytoplasm and differentiate into macronuclear anlagen, and the other 2 of which are located in the anterior region of the cytoplasm and remain as micronuclei; the old macronucleus is broken into fragments (the pattern of nuclear differentiation differs from species to species; see review by Vivier 1974; Wichtermann 1986).

The fertilized nucleus is not necessarily destined to differentiate into the macronucleus, because it behaves as an ordinary germ micronucleus if transplanted into a vegetative cell (Harumoto et al. 1982). What is special is the cytoplasm around this time, because every product of a fertilized nucleus, and even the micronucleus of a vegetative cell, can differentiate into a macronucleus if transplanted into the posterior region of the specific cytoplasm (Mikami 1980; Grandchamp et al. 1981; Mikami et al. 1983b). The old macronucleus plays an important role in creating the specific cytoplasm, because no differentiation into macronuclear anlagen occurs if the old macronucleus is eliminated soon after the first division of the fertilized nucleus (Mikami 1980).

The hypothesis claiming that the time-space-specific cytoplasmic field created in the pre-development cell plays a key role in rewinding the cellular clock can be applied to explain the puzzling question of how the cloned sheep Dolly has overcome the Hayflick limit: in the experiment to create Dolly, the nucleus from an udder cell of a ewe was placed in an unfertilized egg (Wilmut et al. 1997), which is thought to be the spatio-temporally specific cytoplasm for clock rewinding. The signal to start development is then food supply in ciliates and electric pulse in the case of Dolly. In this context, it will be worth studying whether (and to what extent) exconjugant (or exautogamous) cells carrying a macronucleus derived from a transplanted micronucleus will be rejuvenated.

Old macronuclear fragments compete with developing macronuclear anlagen for survival. If fragments survive, thereby defeating anlagen, they can develop into an intact macronucleus after their number is reduced to one per cell by random distribution to daughter cells in the course of post-zygotic cell divisions (the phenomenon known as the MR (macronuclear regeneration)) (Sonneborn 1974b). Nobili (1960) showed that the cell line-age that experienced ten successive MRs had a lifespan similar to (slightly longer than) the progeny derived from normal autogamy, and showed that after each MR the rate of cell division increased temporarily but to a lesser degree with an increasing number of MRs. These results indicate that the old macronucleus was somehow rejuvenated by MR, but its clonal age was not rewound. However, the following possibility still remains: if MR clones were produced selectively from fragments located in the posterior region of ex-autogamous cytoplasm, the clock rewinding could occur.

5.3.3
Clonal Lifespan Versus Cultural Lifespan

When *Paramecium* cells are deprived of food resources, they stop dividing, and the starving cells (quiescent cells) undergo physiological changes called cultural aging (Smith-Sonneborn 1985). The cultural lifespan is referred to the period until starving cells die. The process of cultural aging may be viewed as the cellular response to the environmental stress of starvation, which has been studied extensively in bacteria (Zambrano et al. 1996). It is during this process that paramecia (and also other ciliates) initiate sexual reproduction (meiosis), or, in other words, switch the horizontal genetic axis to the vertical one.

There are some similarities between clonal aging and cultural aging in such respects as increasing morphological and physiological disorder (Elliot et al. 1964; Fok et al. 1981a; Fok et al. 1981b) and increasing UV sensitivity (Yamamoto et al. 1997). The two processes may be more than superficially similar: we recently found that the length of autogamy immaturity became shorter when the new autogamous generation was started from the cells with more advanced cultural age (Ishikawa et al. 1998). However, the relationship between these two processes remains to be defined.

Some relevant information from yeast cells appears to be worth mentioning. Kennedy et al. (1995) isolated mutants of *Saccharomyces cerevisiae* that have long and short cellular lifespans, and found that those with long life-spans were also long in cultural lifespan in the sense that they were more resistant to stresses such as starvation or low temperature than those with short lifespans. The gene responsible for the long cellular lifespan and stress resistance was identified as the silencing gene *SIR4*, the product of which is a member of the protein complex called SIR (silent information receptor).

5.4

Coupling Between the Lifespan and Sexual Reproduction

Although the time-measuring mechanism is still unknown, it appears certain that the life-cycle staging of ciliates is genetically controlled. If so, alterations of the genetic program may alter the timing of the life cycle stages. Isolation of mutants with modified life cycle stages may lead to validation of this idea.

In *P. bursaria,* which includes 4 mating types determined by 2 alleles (mating type I: *AABB, AABb, AaBB, AaBb*; II: *aaBB, aaBb*; III: *aabb*; IV: *AAbb, Aabb*), mature cells of a given mating type can mate with the cells of the other 3 mating types by complementarity between the gene products A and a, and B and b. One of the two alleles (*A-a* allele or *B-b* allele) is expressed first, and expression of the other follows. The lag-time is called adolescence. The adolescent cells can mate with only 2 of the other 3 mating types; for example, adolescent *AaBb* cells (mating type I) expressing *A-a* allele (producing A) can mate with the cells of mating types II and III (both producing a), but can not mate with IV (producing A). The order of which one of the 2 alleles is expressed first is genetically controlled, so that mutants with defects in the expression of the second allele become permanently adolescent expressing the first allele until clonal death (Siegel 1967).

In both *Tetrahymena thermophila* (Bleymann 1971) and *P. caudatum* (Myohara et al. 1978), mutants with short lengths of sexual immaturity have been isolated. They are all semidominant; dominant homozygotes have a shorter immaturity period than heterozygotes (in *P. caudatum,* for example, 20–25 cell divisions in dominant homozygotes, about 35 in heterozygotes, and about 50 in the wild type recessive homozygotes).

In *P. tetraurelia,* a mutant with an extremely short clonal life span (about one-tenth of the wild type) has been isolated in the author's laboratory (Takagi et al. 1987b). Breeding experiments revealed that recessive mutation in a single gene named *jumyo* was responsible (Takagi et al. 1989). It has been shown that some previously isolated mutants of *P. tetraurelia,* such as *am* (abnormal macronuclear distribution to daughter cells) and some groups of *nd* (trichocysts non-discharging) with macronuclear misdivision, have shorter lifespans than that of the wild type (Aufderheide et al. 1985). The *jumyo* mutant is, however, normal in most of the aspects so far examined, including macronuclear division, trichocyst discharging, and the functions involved in sexual reproduction. The survival curve plotted against the clonal age shows a pattern of persistence of 100 % survival followed by a sudden drop, not a linear regression (Fig. 3).

An unusual characteristic found in the *jumyo* mutant is that it can undergo autogamy even when food is abundant. For some time after this mutant was isolated, the clonal termination was defined by either clonal death or autogamy; since then, cells terminating in autogamy have come to dominate the population of the stock culture. Also, in the wild type cells,

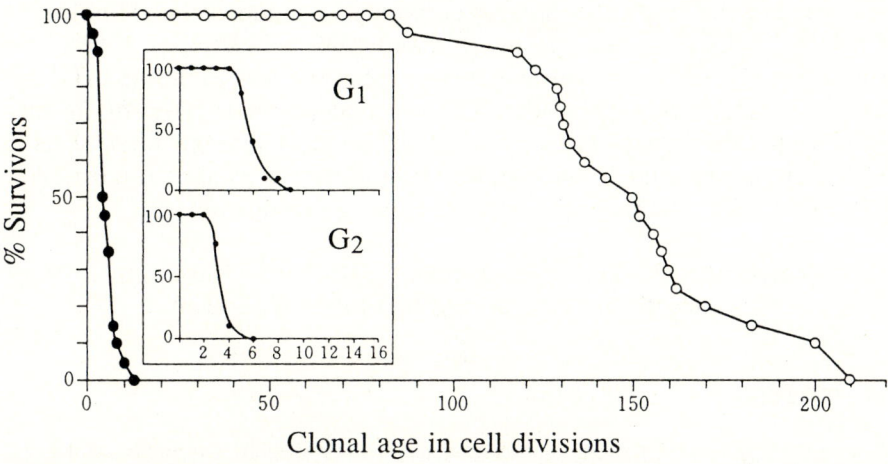

Fig. 3. Survival curves of wild type (*open circles*) and *jumyo* mutant (*closed circles*). Inlet shows enlarged survival curves of the *jumyo* mutant in 2 successive autogamous generations (numbers on the abscissa indicate clonal age in cell divisions

autogamy occurs in nutrient rich conditions at late senescence (Sonneborn 1974b; Takagi et al. 1987a). Clonal termination by autogamy does not practically differ from clonal termination by death, because autogamy at late senescence produces no viable progeny. In the *jumyo* mutant, however, autogamy occurs frequently under nutrient-rich conditions, and at very young clonal age (with short interautogamous intervals) to produce fertile progeny. These observations appear to support the speculation arguing that the sexual reproduction (haploidization system) is basically coupled to the cellular lifespan (brake system for stopping cell division).

Although the somatic cells of multicellular organisms which have cellular lifespans seems to have nothing to do with sexual reproduction, a cryptic relationship between the brake system for suppressing cell division in soma or even for destroying the soma, and the accelerator system for promoting meiosis in germ, sounds natural in ciliates, in which evolutionarily primitive systems are thought to have been preserved more than in mammals. During conjugation and autogamy in ciliates, the germinal micronucleus undergoes meiosis and fertilization, while the somatic macronucleus shrinks or is degraded to fragments and finally disappears. The commitment to macronuclear degradation occurs almost concomitantly with the commitment to micronuclear meiosis (for review, see Fujishima 1988). Note that the life-managing functions in ciliates are carried out by the macronucleus, including the initiation, commitment and prosecution of sexual reproduction (Mikami 1979, 1988). It is noteworthy that the macronuclear degradation is suicidal (apoptotic), because apoptotic DNAs of ~200 bp and multiples thereof are detectable on agarose gel electrophoresis (Davis et al. 1992; Mpoke et al. 1996), and because macronuclear degradation is inhibited by inhibitors of gene expression such as actinomycin D and cycloheximide and by nuclease inhibitors (e.g., aurin) (Mikami 1996; Mpoke et al. 1996). The speculation of cryptic association between the brake system in soma and the accelerator system in germ leads me to speculate further: "Destroy the circuit of sexual reproduction, and you will get an eternal asexual life." This is testable in *Paramecium* by looking for an immortal mutant of *P. tetraurelia* by screening the cells that are unable to undergo autogamy.

Acknowledgments. The author is grateful to Dr. Terue Harumoto of Nara Women's University for her critical reading of the manuscript.

References

Alberts B, Bray D, Lewis J, Raff M, Roberts K, Watson JD (1994) Molecular biology of the cell, third edition. Garland Publishing, New York London

Ameisen JC (1998) The evolutionary origin and role of programmed cell death in single-celled organisms: a new view of executioners, mitochondria, host-pathogen interactions, and the role of death in the process of natural selection. In: Lockshin RA, Zakeri Z, Tilly JL (eds) When cells die. Wiley-Liss, New-York, pp 3–56

Aufderheide KJ (1984) Clonal aging in *Paramecium tetraurelia*. Absence of evidence for cytoplasmic factor. Mech Ageing Dev 28: 57–66

Aufderheide KJ, Schneller M (1985) Phenotypes associated with early clonal death in *Paramecium tetraurelia*. Mech Ageing Dev 32: 299–309

Berger JD (1988) The cell cycle and regulation of cell mass and macronuclear DNA content. In: Görtz H-D (ed) Paramecium. Springer-Verlag, Berlin, pp 97–119

Bleymann LK (1971) Temporal patterns in the ciliated protozoa. In: Cameron IL, Padilla GM, Zimmer AM (eds) Developmental aspects of the cell cycle. Academic Press, New York, pp 67–91

Bodnar AG, Ouellette M, Frolkis M, Holt SE, Chiu C-K, Morin GB, Harley CB, Shay JW, Lichtsteiner S, Wright WE (1998) Extension of life-span by introduction of telomerase into normal human cells. Science 279: 349–352

Bunn CL, Tarrant GM (1980) Limited lifespan in somatic cell hybrids and cybrids. Exp Cell Res 127: 385–396

Cutler RG (1978) Evolutionary biology of senescence. In: Behnke JA, Finch CE, Moment GB (eds) The biology of aging. Plenum Press, New York, pp 311–360

Davis MC, Ward JG, Herrick G, Allis CD (1992) Programmed nuclear death: apoptotic-like degradation of specific nuclei in conjugating *Tetrahymena*. Dev Biol 154: 419–432

Elliot AM, Bak IJ (1964) The fate of mitochondria during aging in *Tetrahymena pyriformis*. J Cell Biol 20: 113–129

Fok AK, Allen RD (1981a) Axenic *Paramecium caudatum*. II. Changes in fine structure with culture age. Eur J Cell Biol 25:182–192

Fok AK, Allen RD, Kaneshiro ES (1981b) Axenic *Paramecium caudatum*. III. Biochemical and physiological changes with culture age. Eur J Cell Biol 25:193–201

Freiburg M (1988) Organization and expression of the nuclear genome. In: Görtz H-D (ed) Paramecium. Springer-Verlag, Berlin, pp 141–154

Fujishima M (1988) Conjugation. In: Görtz H-D (ed) Paramecium. Springer-Verlag, Berlin, pp 70–84

Galadjieff MA, Metalnikov S (1933) L'immortalite de la cellule. Vingtdeux ans de culture d'infusoires sans conjugaison. Archs Zool Exp Gen 75: 331–352

Gilley D, Blackburn EH (1994) Lack of telomere shortening during senescence in *Paramecium*. Proc Natl Acad Sci USA 91: 1955–1958

Grandchamp S, Beisson J (1981) Positional control of nuclear differentiation in *Paramecium*. Dev Biol 81: 336–341

Greider CW (1996) Telomere length regulation. Annu Rev Biochem 65: 337–365

Grell KG (1973) Protozoology. Springer-Verlag, Berlin

Haga N, Hiwatashi K (1981) A protein called immaturin controlling sexual maturity in *Paramecium*. Nature 289: 177–179

Harley CB, Futcher AB, Greider CW (1990) Telomeres shorten during ageing of human fibroblasts. Nature 345: 458–460

Harley CB (1991) Telomere loss: mitotic clock or genetic time bomb? Mutat Res 256: 271–282

Harumoto T, Hiwatashi K (1982) Transplantation of synkaryon in *Paramecium caudatum*. Exp Cell Res 137: 476–481

Hayflick L, Moorhead PS (1961) The serial cultivation of human diploid cell strains. Exp Cell Res 25: 585–621

Iizima S, Numata M, Miwa I. (1997) Carry-over effects on the immaturity of autogamy in *Paramecium tetraurelia*. Zool Sci 14 (Suppl.) : 27

Ishikawa Y, Suzuki A, Takagi Y (1998) Factors controlling the length of autogamy-immaturity in *Paramecium tetraurelia*. Zool Sci 15: 707–712

Karino S, Hiwatashi K (1981) Analysis of germinal aging in *Paramecium caudatum* by micronuclear transplantation. Exp Cell Res 136: 407–415

Karino S, Hiwatashi K (1984) Resistance of germinal nucleus to aging in *Paramecium*: evidence obtained by micronuclear transplantation. Mech Ageing Dev 26: 51–66

Karrer KM (1986) The nuclear DNAs of holotrichous ciliates. In: Gall JG (ed) The molecular biology of ciliated protozoa. Academic Press, Orlando, pp 85–110

Kennedy BK, Austriaco Jr NR, Zhang J, Guarente L (1995) Mutation in the silencing gene *SIR4* can delay aging in *S. cerevisiae*. Cell 80: 485–496

Klass MR, Smith-Sonneborn J (1976) Studies on DNA content, RNA synthesis, and DNA template activity in aging cells of *Paramecium aurelia*. Exp Cell Res 98: 63–72

Klobutchr LA, Prescott DM (1986) The special case in the hypotrichs. In: Gall JG (ed) The molecular biology of ciliated protozoa. Academic Press, Orlando, pp 111–154

Kolter R, Losick R (1998) One for all and all for one. Science 280: 226–227

Kroll RJ, Barnett A (1968) The effect of different fission rates on the onset of maturity in *Paramecium multimicronucleatum*. J Protozool 15 (Suppl): 10

Mikami K (1979) Internuclear control of DNA synthesis in exconjugant cells of *Paramecium caudatum*. Chromosoma 73: 131–142

Mikami K (1980) Differentiation of somatic and germinal nuclei correlated with intracellular localization in *Paramecium caudatum* exconjugants. Dev Biol 80: 46–55

Mikami K (1988) Nuclear dimorphism and function. In: Görtz H-D (ed) Paramecium. Springer-Verlag, Berlin, pp 85–96

Mikami K (1996) Repetitive micronuclear divisions in the absence of macronucleus during conjugation of *Paramecium caudatum*. J Euk Microbiol 43: 43–48

Mikami K, Koizumi S (1983a) Microsurgical analysis of the clonal age and the cell-cycle stage required for the onset of autogamy in *Paramecium tetraurelia*. Dev Biol 100: 127–132

Mikami K, Ng SF (1983b) Nuclear differentiation in *Paramecium tetraurelia*. Transplantation of vegetative micronuclei into early exconjugants. Exp Cell Res 144: 25–30

Miwa I (1984) Destruction of immaturin activity in early mature mutants of *Paramecium caudatum*. J Cell Sci 72: 111–120

Miwa I, Hiwatashi K (1970) Effect of mitomycin C on the expression of mating ability in *Paramecium caudatum*. Jpn J Genet 45: 269–275

Miwa I, Haga N, Hiwatashi K (1975) Immaturity substances: material basis for immaturity in *Paramecium*. J Cell Sci 19: 369–378

Miyake A, Rivola V, Harumoto T (1991) Double paths of macronucleus differentiation at conjugation in *Blepharisma japonicum*. Europ J Protistol 27: 178–200

Mpoke S, Wolfe J (1996) DNA digestion and chromatin condensation during nuclear death in *Tetrahymena*. Exp Cell Res 225: 3357–3365

Muggleton-Harris AL, De Simone DW (1980) Replicative potentials of various fusion products between WI-38 and SV 40 transformed WI-38 cells and their components. Somatic Cell Genet 6: 689–698

Myohara K, Hiwatashi K (1978) Mutants of sexual maturity in *Paramecium caudatum* selected by erythromycin resistance. Genetics 90; 227–241

Nanney D (1974) Aging and long-term temporal regulation in ciliated protozoa: a critical review. Mech Ageing Dev 3: 81–105

Nanney D (1980) Experimental ciliatology. John Wiley & Sons, New York

Nobili R (1960) The effect of macronuclear regeneration on vitality in *Paramecium aurelia*, syngen 4. J Protozool 7(Suppl): 15

Olovnikov AM (1973) A theory of marginotomy. The incomplete copying of template margin in enzymic synthesis of polynucleotides and biological significance of the phenomenon. J Theor Biol 41: 181–190

Orias E (1986) Ciliate conjugation. In: Gall JG (ed) The molecular biology of ciliated protozoa. Academic Press, Orlando, pp 45–84

Orias E (1998) Mapping the germ-line and somatic genomes of a ciliated protozoan, *Tetrahymena thermophila*. Genome Res 8: 91–99

Pereira-Smith OM, Smith JR (1983) Evidence for the recessive nature of cellular immortality. Science 221: 964–966

Raikov IB (1982) The protozoan nucleus: morphology and evolution. Springer-Verlag, New York

Raikov IB (1995) Structure and genetic organization of the polyploid macronucleus of ciliates: a comparative review. Acta Protozool 34: 151–171

Schwartz V, Meister H (1973) Eine Altersveränderung des Makronucleus von *Paramecium*. Z Naturforsch 28c: 232

Schwartz V, Meister H (1975) Einige quantitative Daten zum Problem des Alterns bei *Paramecium*. Arch Protistenk Biol 117: 85–109

Siegel RW (1961) Nuclear differentiation and transitional cellular phenotypes in the life cycle of *Paramecium*. Exp Cell Res 24: 6–20

Siegel RW (1967) Genetics of ageing and the life cycle in ciliates. Symp Soc Exp Biol 21: 127–148

Smith-Sonneborn J (1971) Age correlated sensitivity to ultraviolet radiation in *Paramecium*. Radiat Res 46: 64–69

Smith-Sonneborn J (1979) DNA repair and longevity assurance in *Paramecium tetraurelia*. Science 203: 1115–1117

Smith-Sonneborn J (1981) Genetics and aging in protozoa. Int Rev Cytol 73: 319–354

Smith-Sonneborn J (1985) Aging in unicellular organisms. In: Finch CE, Schneider EL (eds) Handbook of the biology of aging, second edition. Van Nostrand Reinhold, New York, pp 79–104

Smith-Sonneborn J, Klass M, Cotton D (1974) Parental age and life-span versus progeny life-span in *Paramecium*. J Cell Sci 14: 691–699

Smith-Sonneborn J, Reed JC (1976) Calendar life-span versus fission life-span of *Paramecium aurelia*. J Gerontol 331: 2–7

Sonneborn TM (1954) The relation of autogamy to senescence and rejuvenescence in *Paramecium aurelia*. J Protozool 1: 38–53

Sonneborn TM (1974a) *Tetrahymena pyriformis*. In: Mayr E (ed) Handbook of genetics, vol 2. Plenum, New York London, pp 433–467

Sonneborn TM (1974b) *Paramecium aurelia*. In: Mayr E (ed) Handbook of genetics, vol 2. Plenum, New York London, pp 469–594

Sonneborn TM, Schneller MV (1960a) Physiological basis of aging in *Paramecium*. In: Strehler BL (ed) The biology of aging. Waverly Press, Baltimore, pp 283–284

Sonneborn TM, Schneller MV (1960b) Age-induced mutations in *Paramecium*. In: Strehler BL (ed) The biology of aging. Waverly Press, Baltimore, pp 286–287

Sonneborn TM, Schneller MV (1960c) Measures of the rate and amount of aging on the cellular level. In: Strehler BL (ed) The biology of aging. Waverly Press, Baltimore, pp 290–291

Takagi Y (1970) Expression of the mating-type trait in the clonal life history after conjugation in *Paramecium multimicronucleatum* and *Paramecium caudatum*. Jpn J Genet 45: 11–21

Takagi Y (1988) Aging. In: Görtz H-D (ed) Paramecium. Springer-Verlag, Berlin, pp 131–140

Takagi Y (1993) The life-span of cells and of organisms: from a viewpoint of *Paramecium* (in Japanese), Heibonsya, Tokyo

Takagi Y, Yoshida M (1980) Clonal death associated with the number of fissions in *Paramecium caudatum*. J Cell Sci 41: 177–191

Takagi Y, Kanazawa N (1982) Age-associated changes in macronuclear DNA content in *Paramecium caudatum*. J Cell Sci 54: 137–147

Takagi Y, Nobuoka T, Doi M (1987a) Clonal lifespan of *Paramecium tetraurelia*: effect of selection on its extension and use of fissions for its determination. J Cell Sci 88: 129–138

Takagi Y, Suzuki T, Shimada C (1987b) Isolation of a *Paramecium tetraurelia* mutant with short clonal life-span and with novel life-cycle features. Zool Sci 4: 73–80

Takagi Y, Izumi K, Kinoshita H, Yamada T, Kaji K, Tanabe H (1989) Identification of a gene that shortens clonal life span of *Paramecium tetraurelia*. Genetics 123: 749–754

Thomas J, Nyberg D (1988) Vitamin E supplementation and intense selection increase clonal life span in *Paramecium tetraurelia*. Exp Gerontol 23 : 501–512

Vivier E (1974) Morphology, taxonomy and general biology of the genus *Paramecium*. In: Van Wagtendonk WJ (ed) Paramecium. A current survey. Elsevier, Amsterdam, pp 1–89

Wichterman R (1986) The biology of paramecium, Second edition. Plenum Press, New York

Wilmut I, Schnieke AE, McWhir J, Kind AJ, Campbell HS (1997) Viable offspring derived from fetal and adult mammalian cells. Nature 385: 810–813

Yamamoto N, Hayashihara, Takagi Y (1997) Changes in UV sensitivity with cell cycle, clonal age, and cultural age in *Paramecium tetraurelia*. Zool Sci 14: 747–752

Zambrano MM, Kolter R (1996) GASPing for life in stationary phase. Cell 86: 181–184

Zambrano MM, Siegele DA, Almiron M, Tormo A, Kolter R (1993) Microbial competition: *Escherichia coli* mutants that take over stationary phase cultures. Science 259: 1757–1760

Cellular Genealogy of In-Vitro Senescence and Immortalization

T. Matsumura[1]

1
Introduction

1.1
In Vitro Senescence and Immortalization of Somatic Cell Populations

A population of mammalian somatic cells in culture in most cases terminates its proliferation after a period of vigorous growth. Occasionally, either through the action of extraneous effectors or spontaneously, it becomes immortalized, resulting in a continuously growing cell population (Hayflick and Moorhead 1961; Todaro and Green 1963).

The finite proliferative life of a cultured somatic cell population (in vitro senescence) and its transition to an indefinitely proliferative state (immortalization) have been extensively studied biochemically at the cellular and molecular levels, providing fundamental knowledge on the mechanisms of aging and cancer causation (Hayflick 1977; Harley 1997).

Human cells and chicken cells are rarely immortalized without assistance of viral oncogenes, such as SV-40 T antigen gene, irradiation, or chemical carcinogens (Girardi et al. 1965; Namba et al. 1978; McCormick and Maher 1988;), while rodent cells are frequently immortalized spontaneously without any extraneous effectors (Todaro and Green 1963; Macieira-Coelho and Azzarone 1988). Primate and avian cells show relatively long in vitro proliferative life spans compared with rodent cells. Nevertheless, mammalian and avian species share a number of common features in their in vitro cellular proliferation.

Among numerous biomarkers and genetic traits examined, the finding that the length of telomere repeat at chromosomal termini shortens during the finite proliferative life span of human diploid cells while it is maintained in a number of cancer cells has led to the telomere hypothesis of aging and immortalization (Harley et al. 1990; Hiyama, et al. 1995; Harley 1997; Sedivy 1998). The presence of four cytogenetically complementary groups of immortalized cells, of which a fusion of two cells from two different comple-

---

[1] Meiji Cell Technology Center, Meiji Milk Products Co., Ltd. 540 Naruda, Odawara 250–0862, Japan.

Progress in Molecular and Subcellular Biology, Vol. 24
A. Macieira-Coelho
© Springer-Verlag Berlin Heidelberg 1999

mentary groups gives rise to a finitely proliferative cell population, is another important finding (Pereira-Smith and Smith 1988).

Although biomarkers of senescence and immortalization have been ascertained, a unifying concept of senescence and immortalization has yet to be established. The fact that many immortalized cell populations contain finite proliferative cells (Martin et al. 1974; Smith and Whitney 1980), while biomarkers are obtained from whole cell populations, makes such studies difficult. On the other hand, it is theoretically difficult to reject the possibility that a cell culture population of finite proliferative life span contains indefinitely proliferative cells (Holliday 1981).

To unify the concept of senescence and immortalization, it is therefore essential to take the proliferative characteristics of individual cells into account in interpreting the biomarkers of the respective cell populations.

1.2
Genealogical Approach

A number of approaches have been introduced to analyze the heterogeneity of proliferation parameters. The heterogeneity during the interphase period can be quantified from radioautograms of cells continuously labeled with tritiated thymidine (Cristofalo and Sharf 1973; Macieira-Coelho 1974; Matsumura et al. 1979a). A histogram of the number of cells in individual colonies grown in a culture dish is also an indicator of the heterogeneity of colony-forming capability, or 'clonogenicity', of individual cells (Martin et al. 1974). Furthermore, with a routine transfer protocol for cell culture, the distribution of the proliferative life span among the individual colonies can be determined (Smith and Whitney 1980). Probably, cell family trees provide the most detailed information of the structure of proliferative heterogeneity: they provide not only quantitative parameters of cell proliferation, but also information about relationships among cells.

The cell genealogy of finitely proliferative life spans was first studied by Absher and coworkers using human diploid cells (HDCs; Absher et al. 1974; Absher and Absher 1975). With HDCs, however, the chance of covering the process of spontaneous immortalization on film was virtually zero.

A rodent cell culture system shows some advantages over HDCs in genealogical studies, since it senesces in a few weeks, in contrast to several months with HDCs. Occasionally it becomes immortalized spontaneously, offering the possibility of covering the whole process of immortalization as well as senescence in cell family trees. The use of epithelial-like cells rather than fibroblastic cells is also advantageous because the former move much more slowly than the latter, making film analyses easy.

With the above background, we initiated a genealogical study using primary and serially transferred cultures of rat liver epithelial-like cells (RLECs; Matsumura 1983, 1984; Matsumura et al. 1983, 1985a). To our knowledge, family tree analyses of senescence and immortalization have not been done since then. On the other hand, molecular studies based on cell populations have advanced greatly during the last 10 years (Harley 1997; Sedivy 1998). The purpose of this chapter is first to summarize the general genealogical features of in vitro senescence and immortalization, and then to discuss their mechanisms using the bases of cellular genealogy.

2
Methods of Cellular Genealogy

2.1
Obtaining a Cell Family Tree

An experiment to obtain cell family trees is initiated by setting a culture flask inoculated with a small number of cells on an observation plate of an inverted phase-contrast microscope with a warming chamber, choosing a colony of cells within a microscopic field, and starting either time-lapse cinemicrography or video recording.

A detail which may need some expertise is the periodic reduction of the number of cells in a culture. This can be attained by marking the position of an observation field on the outer surface of a culture flask so that it can be reset exactly to the position after manipulation, then taking the culture flask into a safety cabinet, removing cells around the field of observation by using a fine needle under a microscope, renewing the tissue culture medium, and then returning the flask back to the microscopic observation table.

A chart of a cell family tree can be constructed from either a film or a recording tape simply by tracing all cells generated from one cell followed throughout the period of cinemicrographic observation. Detailed methods for RLECs have been described previously (Matsumura et al. 1983).

2.2
Analyzing a Cell Family Tree

As for many other cells, the following cellular events are observed in RLECs (Matsumura et al. 1983):

Mitosis. This is easily noted by the contraction of cytoplasm and subsequent formation of two daughter cells.

Incomplete Mitosis. This is determined by cell-rounding up (cytoplasmic contraction) and cell-splitting immediately followed by the fusion of the two daughter cells to form either a cell with a single polyploid nucleus or a binu-cleated (or multinucleated) cell.

Cell Death. This is determined by an abrupt ceasing of cell movement which is usually preceded by bubbling, and followed by shrinking of the cytoplasm.

Escaping From Sight. Because of the limited area of the microscopic observation field, a cell often escapes from sight. A plausible assumption for those cells with unknown terminal events is that they follow the same fates as those remaining in the field. This assumption is made in every statistical analysis in our studies described below.

Interphase cells on a family tree chart are classified into the following three classes (Matsumura 1984) :

1. D cells: those initiated with mitosis and terminated with mitosis (dividing cells).
2. I cells: those initiated with mitosis and terminated with incomplete mitosis (incompletely dividing cells). Although rare, some cells undertake incomplete mitosis again after an incomplete mitosis. They are included in the I cell class, too.
3. N cells: those initiated either with mitosis or with incomplete mitosis and terminated with death (non dividing cells).

Out of all cells present at one time in a family tree, the proportion of those remaining without cell division can be determined as a function of time. This value can be compared with the proportion of non-labeled cells in a radioautogram after exposure of a cell population to tritiated thymidine .

The histogram of the number of cells produced during a fixed period of time from each of their ancestor cells on a family tree chart can be compared with the histogram of the clonogenicity determined by colony formation .

A clonal proliferation profile can be reconstituted for any one cell in a family tree. Such a profile can be compared with a clonal proliferation profile obtained by actual cell cloning and repeated culture transfers.

In addition, such relative parameters as the interphase time length of a mother cell versus that of a daughter cell, or of a pair of sibling cells, can be obtained from family tree charts. The methods of analysis are briefly described in the legend of Fig. 1 (See Matsumura 1984 for details).

3
Genealogy of Finite Proliferation

3.1
Intermitotic Period Distribution

It has been well known that the interphase period varies in a population of HDCs and that the cell cycle time is more heterogeneous in a 'senescent' culture than in a 'young' culture (Absher et al. 1974; Macieira-Coelho 1974; Grove and Cristofalo 1976).

From the family tree analyses of finitely growing RLECs, the semilogarithmic presentation of the cumulative intermitotic period distribution for D cells, known as the α plot, was found to be linear (Matsumura 1984). This indicates that D cells fit with the transition probability model in that a cell

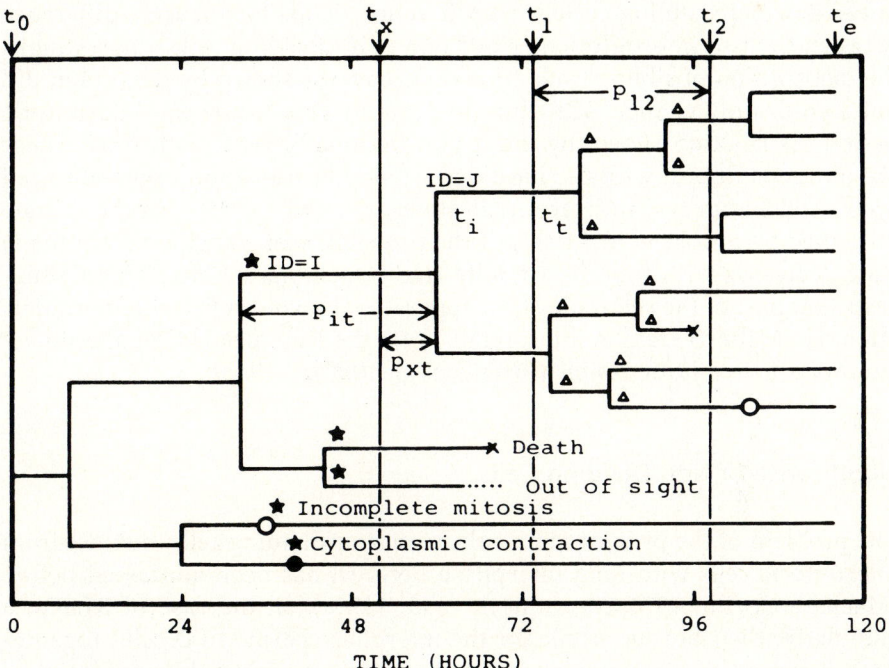

Fig. 1. Definition of quantitative values determined on a family tree. Events identified on films were mitosis (—▢—), incomplete mitosis (—○—), cytoplasmic contraction (—●—), death (—✗—), and moving out of sight (— – – –). Incomplete mitosis and cytoplasmic contraction have been described previously (Matsumura et al. 1983) and are collectively referred to in this paper as incomplete division. For a family tree, 'the cohort of cells on time tx' is defined as consisting of cells present at time tx in the family tree, as shown by cells marked ★ in this family tree. For a cell in the cohort of cells at time tx, as indicated by ID=I in this figure, the interphase time period, and the time period for the cell to remain in interphase after tx, are determined, as indicated by pit and pxt, respectively. All cells in a family tree are regarded as consisting of a set of cells. 'The subset of cells on p12' is defined as consisting of cells generated during the period between times t1 and t2, from all cells in a given family tree, as shown by cells marked ★ in this family tree. For each cell in a cell cohort, the number of descendant cells generated after a given time is enumerated: for the cell indicated by ID=J, which is in the cohort on t1, the number of descendant cells produced during the period p12 is three in this figure. A cumulative distribution profile of the interphase period (the α plot) is obtained from the values of interphase periods determined for a given cell cohort or for a given cell subset. A distribution profile of clonogenicity is obtained from the number of descendant cells for a given cell cohort. The distribution profile at a given moment of the time in which cells remain in interphase can be compared with radio-autograms with continuous tritiated thymidine exposure. (Reproduced from Matsumura 1984 by permission of Wiley-Liss, Inc., a subsidary of John Wiley & Sons, Inc.)

cycle time is composed of the probabilistic A state and the deterministic B phase (Smith and Martin 1973), confirming the fitting of intermitotic cells of 'young' HDCs to the model (Grove et al. 1976).

In the finite RLECs, the semilogarithmic presentation of the cumulative distribution of sibling cycle time difference, i.e. the β plot, was shown to be linear too. Here, sibling cycle time difference stands for the time difference between the two intermitotic periods of a pair of sibling cells. Interestingly, the distribution of sibling cycle time difference, as shown by the β plot, did not significantly change with time in a family tree, while the intermitotic period distribution shown by the α plot became broader with time. These observations fit best with the modified version of transition probability cell cycle model with two indeterminate states, i.e., the A state and the Q state presented by Brooks et al. (1980), if the probability of transition from the Q state decreases with time in a family tree (Matsumura 1984). This fit suggests that part of the cell replication apparatus is involved in limited proliferation, since the Q state is interpreted to be the time needed for the nucleation of a new mitotic center (Brooks et al. 1980).

3.2
Significance of Non-Dividing Cells

The problem of the presence in a culture of non-dividing cells, distinct from intermitotic cells with long interphase periods, has been addressed before (Macieira-Coelho 1974; Matsumura et al. 1979a). In the case of RLECs, a cumulative distribution profile for the interphase period of D cells, together with those cells surviving for more than 7 days with unknown terminal events in family trees, showed a skewed line on a semilogarithmic graph (Fig. 3). On the other hand, as described, a cumulative distribution profile of the interphase period for D cells without the addition of those long survivors shows a straight semilogarithmic line, predicting that the frequency of intermitotic cells with more than a 7 day interphase period will be less than 1 % of total intermitotic cells (Fig. 2). The interphase period for I cells is distributed logarithmically, but is narrower than that for D cells (Matsumura 1984). Therefore, if one takes the transition probability model as granted for both D cells and I cells, it can be predicted that the chance of those interphase cells surviving for more than 7 days to enter into mitosis or incomplete mitosis later is negligibly small. On the other hand, the cumulative distribution profile of interphase periods for those long surviving cells, together with dying cells, showed a logarithmically linear line, supporting our classification in which these long survivors and dying cells together make up one class of cells (class N; Matsumura 1984).

In the analyses of sibling relationships, a statistically significant relationship was noted in RLECs so that a pair of sibling cells tend to have the same fate, whether it is mitosis, incomplete mitosis or death. For example, if one cell dies, then its sibling cell dies with a significantly high probability. The implication of such relationship is that the fate of siblings has already been determined in their mother cell. The observation of uneven distribution of DNA between two siblings at the terminal cell division should be taken into account for the fate of sibling pairs (Macieira-Coelho 1995).

Fig. 2. Schematic presentation of cumulative intermitotic time distribution (α), and a cumulative distribution of sibling cycle time difference (β) redrawn from a family tree of a finitely proliferative RLEC. (For actual data, refer to Fig. 6, Matsumura 1984)

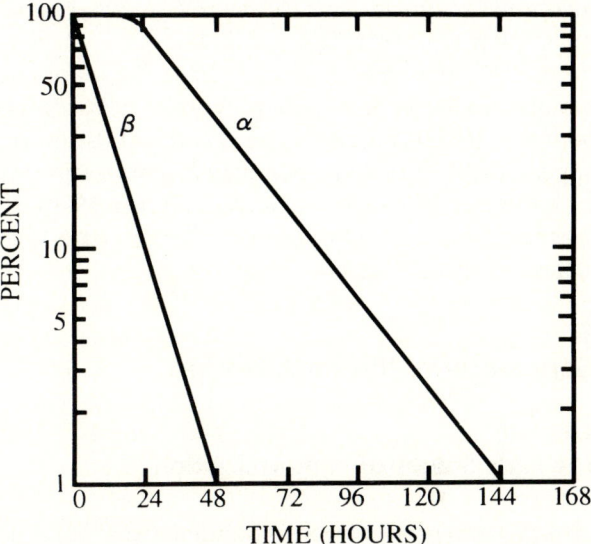

Fig. 3. Schematic presentation of cumulative interphase time distribution profiles from a finitely proliferative RLEC family tree. $\gamma 1$ and $\gamma 2$ for all interphase cells including dying cells and cells surviving for 7 days or more from an early stage ($\gamma 1$) and a late stage ($\gamma 2$) of a family tree, δ for dying cells + cells surviving for 7 days or more. (For actual data, refer to Figs. 5 and 8, Matsumura 1984)

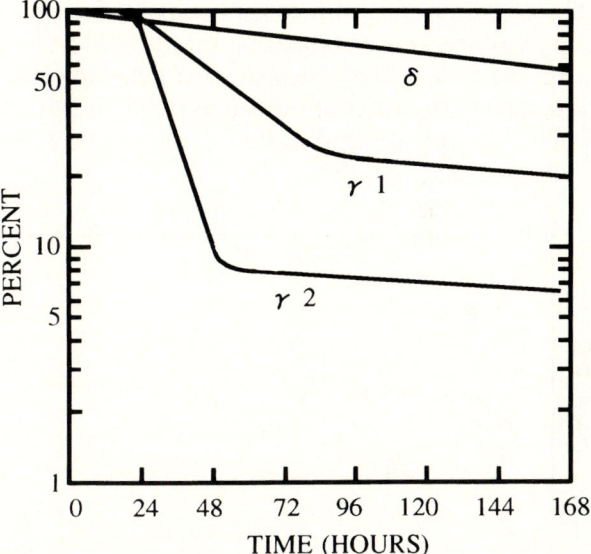

3.3
Significance of Polyploid Cells

Polyploid cells and multinucleated cells often appear in a senescent population of chicken fibroblasts and HDCs during in vitro senescence (Kaji and Matsuo 1979; Matsumura et al. 1979b). RLECs also contain polyploid cells. In the family trees of RLECs, a polyploid cell is always formed after an incomplete mitosis. No cases were seen where a complete mitosis follows an incomplete mitosis (Matsumura 1984).

In HDCs maintained in late phase III, we noted that multinucleated cells, mainly binucleated, and/or polyploid cells comprise almost half of the cell population (Matsumura et al. 1979b; Matsumura 1980). HDCs and RLECs appear to be directed toward polyploidization and multinucleation at the end of the cell cycle life span (Matsumura 1980). It should be noted here that these terminal cells have more to do with pathology than physiological aging (Macieira-Coelho 1995).

4
Genealogy of Immortalization

4.1
The Early Stages of Immortalization

Proliferative profiles were obtained from several family trees of RLECs in primary culture (Matsumura et al. 1985a). Some of them showed limited proliferation, others showed a decline and then a recurrence of proliferation (Fig. 4). No profiles with continuous proliferation from the beginning of primary culture were obtained.

In one family tree which showed a decline and a recurrence of proliferation, an abrupt burst of mitosis was noted (Fig. 5a; Matsumura 1983). This family tree consisted of finitely proliferative cells except for this particular

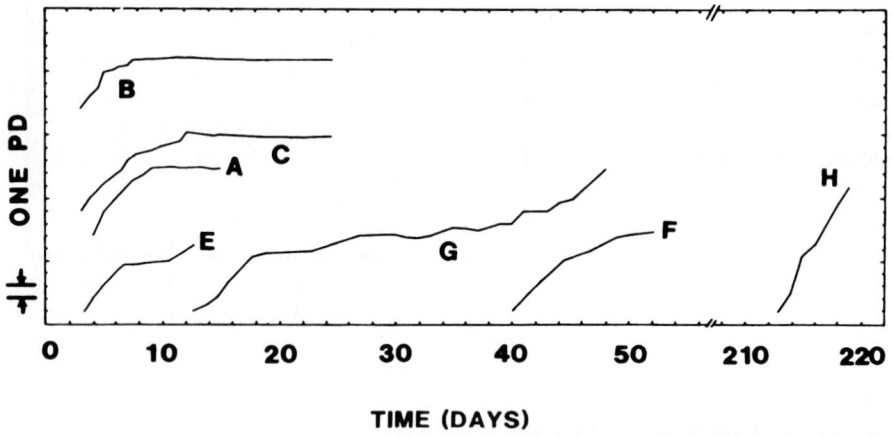

Fig. 4. Poliferative profiles reconstituted from RLEC family trees in either primary (*A, B, C, E, G*) or early passage (*F, H*) cultures. *Letters* in the figure identify family trees. Days on the abscissa are the days of cultivation from the initiation of primary culture. Ordinate scale shows relative values of population doublings (*PD*). The distance between *two bars* shown on the ordinate, indicated by *upward* and *downward arrows*, indicates one population doubling. (Partly reproduced from Fig. 1, Matsumura et al. 1985 by permission of Oxford University Press.)

cell and its progenies, strongly suggesting that it covered the onset of immortalization. Another family tree, also presenting a decline and a recurrence of proliferation, showed a long period of unstable proliferation where heterogeneity in intermitotic periods, non-dividing cells, incompletely dividing cells and dying cells were apparent (Fig. 5b).

Primary cultures with large proliferative colonies were subjected to a routine transfer protocol, and family trees and proliferative profiles were further obtained from these serially transferred cultures (Fig. 4). Since most of the cells, after a burst, underwent finite proliferation at an early stage of immortalization, the mitotic burst may be regarded as due to uncontrolled rejuvenation. In the early stages of immortalization, the cumulative distribution profile of intermitotic time, i.e. the α plot, is semilogarithmically linear, similar to those from finitely proliferative family trees, indicating that the intermitotic time follows the transition probability model (Matsumura et al. 1985a).

4.2
The Advanced Stages of Immortalization

Within 2 months of in vitro culture, however, a serially transferred culture becomes vigorously proliferative, and a family tree obtained in that advanced stage shows homogeneous intermitotic cells with few cells of other classes (Fig. 5c; Matsumura et al. 1985a). The α plot and the β plot for these cells did not show straight lines, indicating that they did no longer fit with the transition probability model.

Previously, RLECs were considered to be immortal diploid cells from the beginning of culture (Tokiwa et al. 1979; Schaefer 1980). A serum-free culture of mouse embryo cells was reported to be immortal from the beginning of culture (Loo et al. 1987). I believe that these previous observations should be very carefully reevaluated with respect to genealogy, since in rodent cells, as mentioned here, a multi-step process of immortalization could happen within a short period of time (Fig. 6c).

Multiple step changes during immortalization have been observed not only for RLECs but for a number of other cell systems, although they do not all evolve the same way. In some established mouse cell lines, unlike established RLEC lines, the intermitotic time distribution still follows the transition model (Shields and Smith 1977). In some established human cell lines, again unlike RLEC lines, heterogeneity in clonogenicity is still maintained (Martines et al. 1978; Matsumura et al. 1985b).

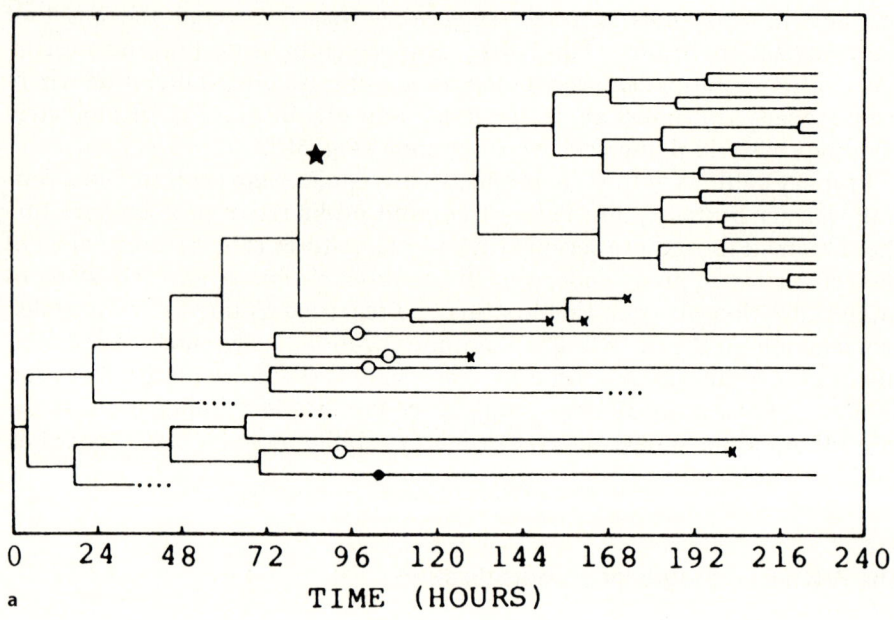

a

TIME (HOURS)

0   24   48   72   96   120   144   168   192   216   240

b

TIME(DAYS)

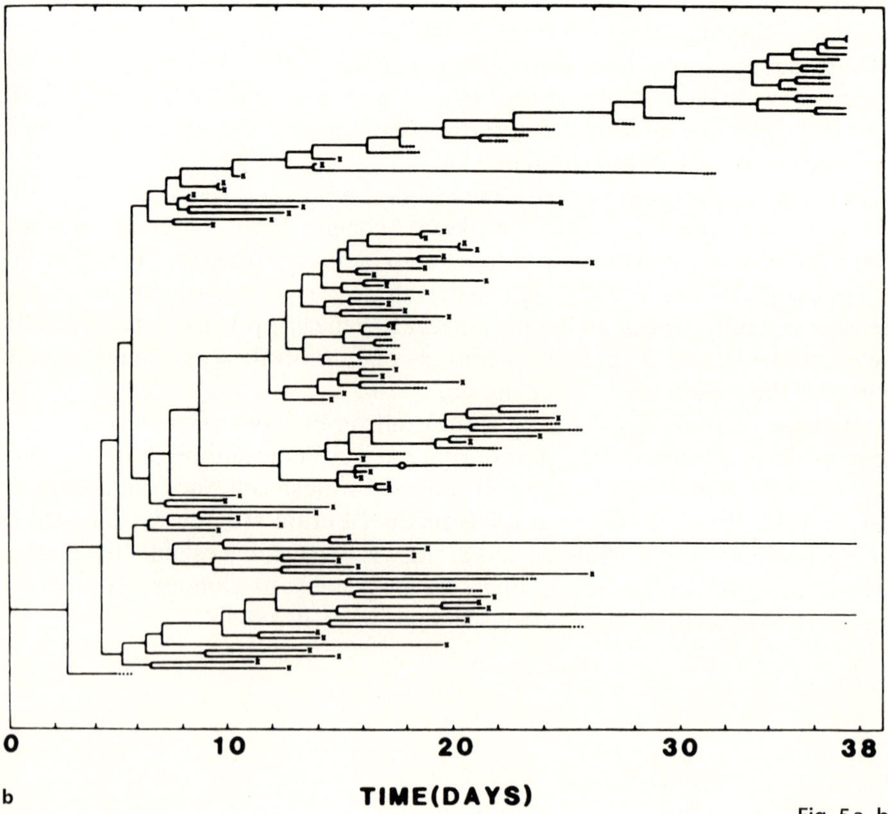

0   10   20   30   38

Fig. 5 a, b.

Fig. 5. Some RLEC family trees representing the process of immortalization. **a.** Family tree *E* showing an abrupt burst of mitosis in a primary culture. **b.** Family tree *G* showing the period of unstable proliferation toward stable growth. **c.** Family tree *H* showing stable proliferation. As described in Fig. 1, events on a family tree were mitosis (—⊏), incomplete mitosis (—○—), cytoplasmic contraction (—●—), death (—✕), and moving out of sight (— – – –). The cell marked ★ in Fig. 5a underwent cell division during the course of the observation period, which is described as 'a burst of mitosis' in the text. As described in the text, cells in such a burst do not continue uniform cell division, but their majority are finitely proliferative. The *arrow* on the abscissa in Fig. 5c shows the time when a part of cells under microscopic observation field were removed.

For proliferative profiles, refer to those denoted *E, G,* and *H* in Fig. 4. (Partly taken from Fig. 1, Matsumura, 1983 and Figs. 3 and 5, Matsumura et al. 1985 by permission of Oxford University Press.)

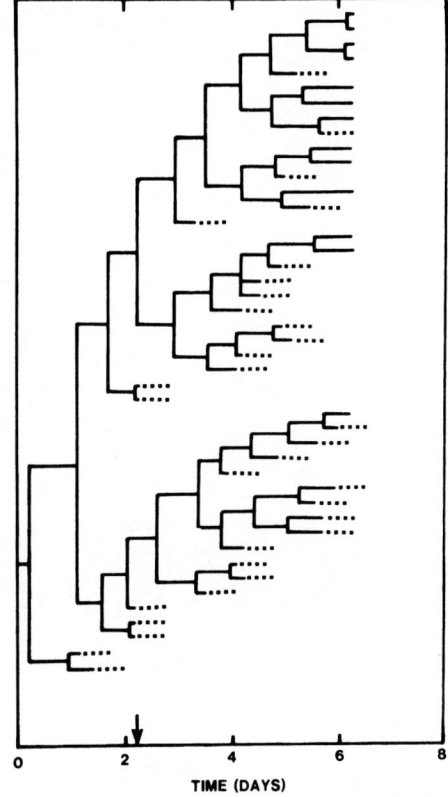

TIME (DAYS)

Fig. 5 c

5
Discussion

5.1
Finite Proliferation Versus Indefinite Proliferation

Before the cell genealogy studies, the essential part of the genealogical characteristics of finitely and indefinitely proliferative populations had already been outlined by radioautographic and colony formation experiments (Cristofalo and Sharf 1973; Macieira-Coelho 1974; Martin et al. 1974; Matsumura et al. 1979a; Smith and Whitney 1980). Conclusions drawn from these previous studies using human cells together with genealogical studies using RLECs are schematically presented in Fig. 6, and summarized as follows:

In both the human and rat cells, a growing finitely proliferative population and an indefinitely proliferative population cannot be distinguished from each other when they are compared with a histogram of clonogenicity:

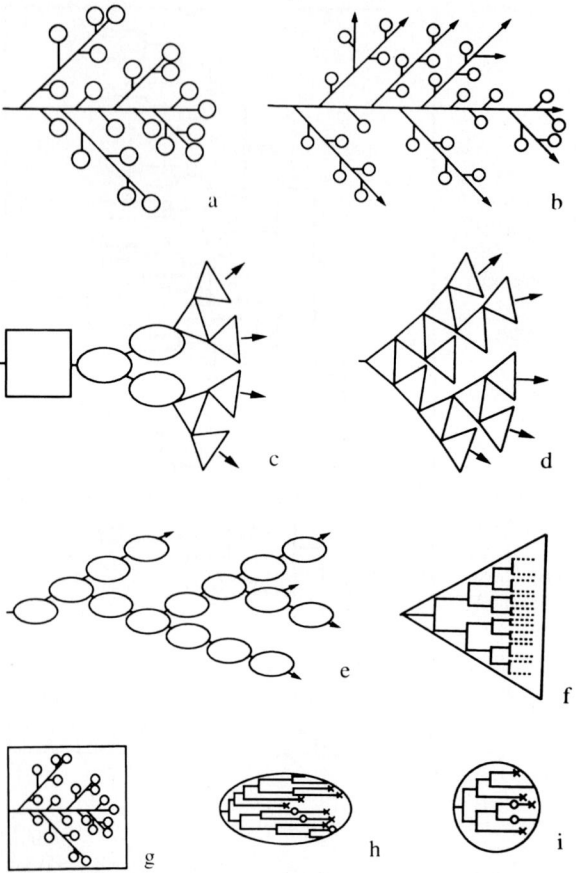

Fig. 6. Schematic presentation of putative genealogies with finite and indefinite cell proliferation. The *triangle, square, oval,* and *round* elements in the schemes **a** to **e** are shown in detail in the schemes **f** to **i**, respectively. Events on a family tree were mitosis (—⊏), incomplete mitosis (—○—), death (—✕), and moving out of sight (— – – –). **a.** Partially clonogenic finite state: This tree-like shape with branched lines and lief-like circles reprents a cell family tree with its initial cell at the left-end. Branched lines represent the part of cell family tree in which cells divide and produce stem cell-like progenies with large, but varying levels of limited proliferative potential. A circle at the end of every branch represents an element of family tree composed of small doubling potential (Refer to Fig. 6. i). **b.** Partially clonogenic indefinite state: This vein-like shape with endlessly branched lines and lief-like circles represent a cell family tree which is the same as Fig. 6. a., except that the proliferative potential of stem-like progenies is unlimited. **c.** An example of multistep process of immortalization: A cell family tree shown here initiates with a cell in a partially clonogenic finite state (Figs. 6a, 6g), through a partially clonogenic intermittent state (Figs. 6e, 6h), toward a totally clonogenic indefinite state (Figs. 6d, 6f) as its ultimate goal. Other multistep processes may be possible. **d.** Totally clonogenic indefinite state: All cells in a population is indefinitely proliferative (Refer to Figs. 5 c, 6d and 6f). **e.** Intermittently clonogenic indefinite state: Occasional rejuvenation of proliferative potential happens in finite proliferative clonogenic cells. Otherwise it is the same as Fig. 6a. **f.** An element of family tree composed of totally clonogenic cells. **g.** An element of family tree composed of partially clonogenic finite cells. **h.** An element of family tree composed of rejuvenated finitely proliferative cells. **i.** An element of family tree composed of small doubling potential

in both cell populations, highly clonogenic cells and cells with poor clonogenicity are separately present, and the poorly clonogenic cells are produced from the highly clonogenic cells stochastically (Martin et al. 1974; Martines et al. 1978; Smith and Whitney 1980; Matsumura 1984; Matsumura et al. 1985a). The number of cell divisions attained by poorly clonogenic cells is around seven for HDCs (Smith and Whitney 1980), while it is only one or two for rat cells (Matsumura 1984).

The only difference noted in colony formation studies between the finite population and the immortalized population, in both the human cell system and the rat cell system, is that the proliferative potential of the highly clonogenic part of the cell population decreases with cell generations in a finite population, while it is maintained in an immortalized cell population (Figs. 6a, 6b; Martines et al. 1978; Smith and Whitney 1980, Matsumura 1984; Matsumura et al. 1985a).

The family trees of RLECs in their early stages of immortalization may be interpreted as being composed of stem cells with unlimited proliferative potential together with branch cells with limited proliferative potential. However they could also be composed of finitely proliferative stem cells and short lived branch cells, as for a finitely proliferative population, provided that mitotic bursts happen in some cells from time to time during the course of populational growth (Matsumura 1983).

If such a mitotic burst, or an uncontrolled rejuvenation, takes place frequently in a cell population, then unlimited proliferation may easily result without the presence of indefinitely proliferative stem cells (Fig. 6e). An immortalized cell population with this second possible mechanism can be fundamentally different either from embryonal stem cells in which every cell lineage in a population is considered to be immortal, or from partly clonogenic immortal cell populations.

To my knowledge, the question has not been adequately answered whether in human cancer cells, the subpopulation with unlimited proliferative potential is maintained with continuous cell division, or with frequent mitotic bursts. Attempts to establish a genealogical classification of indefinitely proliferative cell populations will be of great value.

5.2
Biochemistry, Cell Genetics and Cell Genealogy

In the last few years it has become recognized that the maintenance mechanisms for telomere length play important roles for the maintenance of cell proliferation (Harley et al. 1990; Bryan et al. 1995; Sharma et al. 1995; Lundblad and Wright 1996; Harley 1997; Bodnar et al. 1998). Many permanent cell lines, including embryonal stem cells, show high telomerase activity, while finitely proliferative HDCs show no or very low activity. Some permanent cell lines, on the other hand, show apparently no telomerase activity but are immortal. In HDC populations, the loss of proliferative potential is accompanied by the loss of a number of telomeric repetitive sequences. On the other hand, direct expansion of the proliferative life span is observed when telomerase is activated artificially in HDCs.

Direct expansion of indefinitely proliferative cells, without any intervening step of immortalization (Fig. 6d), is considered to take place in embryonal stem cells and in their cybrid cells reconstituted from enucleated unfertilized eggs and somatic cell nuclei (Wilmut et al. 1997). On the other hand, as stated above, a direct expansion of proliferative life span without an unstable phase is rarely observed in effector-mediated immortalization of somatic HDCs.

The cellular genealogical study of RLECs indicates that immortality is attained not just by one mechanism. It is highly probable that direct immortalization without any sign of an unstable period of proliferation, and an indirect immortalization with an intermittent unstable period of proliferation are caused by different mechanisms. Also, immortalization leading to a totally clonogenic population, to a partially clonogenic population, and to an intermittently clonogenic population could have different mechanisms respectively. For example, if telomerase is responsible for the recovery of telomere repeats, then the activity may be required only intermittently in a small cell fraction for a population to be immortal. Thus, it is of particular interest to study genealogical characteristics of cell lines with direct expansion of their life spans, of effector-mediated immortalized cell lines, and of cancer cell lines with varying levels of telomerase activity. Also, it is important to study the four recessive complementary groups among human continuous cell lines for the same reason. In addition, it may be worth noting here that some immortal populations of cells manifest a finite proliferative life span after chemical treatment (Katakura et al. 1997). It will be particularly interesting to screen chemical conditions or effectors which bring an indefinitely proliferative population back to a finitely proliferative stage, and to study the intermediate steps between their immortal and their finitely proliferative state in both biochemical and genealogical terms.

6
Conclusion

The family tree analysis of cultured cells has great potential, yet is a little utilized approach toward the understanding of the mechanisms of senescence and immortalization. Cultured rat liver epithelial-like cells undergo senescence and spontaneous immortalization within a few weeks, and thus provide an advantageous model for cellular genealogy. They start their in vitro life with finitely proliferative potential. The first sign of immortalization is shown at a low frequency by a cell in a family tree which undergoes a burst of mitosis, while other cells in the same family tree still remain with finite proliferation. Such a burst then creates clonal cells, most of which are still finitely proliferative. These clonal cells pass though an unstable period, and ultimately lead to a stable proliferative cell population.

Immortality of a cell population is not necessarily unique in terms of cellular genealogy. Cell populations composed of uniformly dividing cells (totally clonogenic immortal), of indefinitely proliferative cells and of cells with limited generations (partially clonogenic immortal), or of finitely proliferative cells with occasional occurrence of rejuvenation back to highly proliferative finite cells (intermittently clonogenic immortal) can all be immortal at the level of the cell population.

Biochemical and genetic observations, such as the presence of both telomerase-active and -inactive immortal cell lines and the presence of four cytogenetically recessive complementary groups among immortal cell lines suggest the presence of more than one mechanisms of immortality. Further studies on the classification of, and the relationships among, proliferative populations in terms of biochemistry, genetics as well as genealogy may provide keys to further understand the mechanisms of senescence and immortalization.

References

Absher PM, Absher RG (1975) Time-lapse cinemicrographic studies of cell division patterns of human diploid fibroblasts (WI-38) during their in vitro lifespan. Adv Exp Med Biol 53:91–105

Absher PM, Absher RG, Barnes WD (1974) Genealogies of clones of diploid fibroblasts. Cinemicrophotographic observations of cell division patterns in relation to population age. Exp Cell Res 88:95–104

Bodnar AG, Ouellette M, Frolkis M, Holt SE, Chiu CP, Morin GB, Harley CB, Shay JW, Lichtsteiner S, Wright WE (1998) Extension of life-span by introduction of telomerase into normal human cells. Science 279:349–352

Brooks RF, Bennett, DC, Smith JA (1980) Mammalian cell cycles need two random transitions. Cell 19:493–504

Bryan TM, Englezou A, Gupta J, Bacchetti S, Reddel RR (1995) Telomere elongation in immortal human cells without detectable telomerase activity. EMBO J 14:4240–4248

Cristofalo VJ, Sharf BB (1973) Cellular senescence and DNA synthesis. Thymidine incorporation as a measure of population age in human diploid cells. Exp Cell Res 76: 419–427

Girardi AJ, Jensen FC, Koprowski Hl (1965) SV40-induced transformation of human diploid cells: crisis and recovery. J Cell Comp Physiol 65:69–83

Grove GL, Cristofalo VJ (1976) The 'transition probability model' and the regulation of proliferation of human diploid cell cultures during aging. Cell Tissue Kinet 9:395–399

Harley CB (1997) Human ageing and telomeres. Ciba Found Symp 211:129–139

Harley CB, Futcher AB, Greider CW (1990) Telomeres shorten during ageing of human fibroblasts. Nature 345:458–460.

Hayflick L (1977) The cellular basis for biological aging. In: Finch CE, Hayflick L (eds) Handbook of the biology of aging. Van Nostrand Reinhold New York, pp 159–186

Hayflick L, Moorhead PS (1961) The serial cultivation of human diploid cell strains. Exp Cell Res 25:585–621

Hiyama E, Yokoyama T, Tatsumoto N, Hiyama K, Imamura Y, Murakami Y, Kodama T, Piatyszek MA, Shay JW, Matsuura Y (1995) Telomerase activity in gastric cancer. Cancer Res 55:3258–3262

Holliday R, Huschtscha LI, Kirkwood TB (1981) Cellular aging: further evidence for the commitment theory. Science 213:1505–1508

Kaji K, Matsuo M (1979) Aging of chick embryo fibroblasts in vitro. . Polyploid cell accumulation. Exp Cell Res 119:231–236

Katakura Y, Yamamoto K, Miyake O, Yasuda Y, Uehara N, Nakata E, Kawamoto S, Shirahata S (1997) Bidirectional regulation of telomerase activity in a subline derived from human lung adenocarcinoma. Biochem Biophys Res Commun 18: 313–317

Loo DT, Fuquay JI, Raqson CL, Barnes DW (1987) Extended culture of mouse embryo cells without senescence: inhibition by serum. Science 236:200–202

Lundblad V, Wright WE (1996) Telomeres and telomerase: a simple picture becomes complex. Cell 87:369–375

Macieira-Coelho A (1974) Are nondividing cells present in aging cell cultures? Nature 248: 421–422

Macieira-Coelho A (1995) The last mitoses of the human fibroblast proliferative life span, physiopathologic implications. Mech Ageing Dev 82:91–104

Macieira-Coelho A, Azzarone B (1988) Transition from primary culture to spontaneous immortalization in mouse fibroblast population. Anticancer Res 8:669–676

Martin GM, Sprague CA, Norwood TH, Pendergrass WR (1974) Clonal selection, attenuation and differentiation in an in vitro model of hyperplasia. Am J Pathol 74:137–153

Martines AO, Norwood TH, Prothero JW, Martin GM (1978) Evidence for clonal attenuation of growth potential in HeLa cells. In Vitro 14:996–1002

Matsumura T (1980) Multinucleation and polyploidization of aging human cells in culture. Adv Exp Med Biol 129:31–38

Matsumura T (1983) A rare family tree of cultured rat cells showing a change in proliferative potential. Cell Biol Int Rep 7:931–935.

Matsumura T (1984) Sequence of cell life phases in a finitely proliferative population of cultured rat cells: a genealogical study. J Cell Physiol 119:145–154

Matsumura T, Pfendt EA, Hayflick L (1979a) DNA synthesis in the human diploid cell strain WI-38 during in vitro aging: an autoradiography study. J Gerontol 34:323–327

Matsumura T, Zerrudo Z, Hayflick L (1979b) Senescent human diploid cells in culture: survival, DNA synthesis and morphology. J Gerontol 34:328–334

Matsumura T, Masuda K, Murakami Y, Konishi R (1983) Family trees representing the finitely proliferative nature of cultured rat liver cells. Cell Struct Funct 8:293–301

Matsumura T, Hayashi M, Konishi R (1985a) Immortalization in culture of rat cells: a genealogic study. J Natl Cancer Inst 74:1223–1232

Matsumura T, Nagata M, Konishi R, Goto M (1985b) Studies of SV40-infected Werner syndrome fibroblasts. In: Salk D, Fujiwara Y, Martin GM (eds) Werner's syndrome and human aging. Plenum Publishing, New York, pp 313–330

McCormick JJ, Maher VM (1988) Towards an understanding of the malignant transformation of diploid human fibroblasts. Mutat Res 199:273–291

Namba M, Nishitani K, Kimoto T (1978) Carcinogenesis in tissue culture. 29: Neoplastic tranformation of a normal human diploid cell strain WI-38 with Co-60 gamma rays. Jpn J Exp Med 48:303–311

Pereira-Smith OM, Smith JR (1988) Genetic analysis of indefinite division in human cells: identification of four complementation groups. Proc Natl Acad Sci USA 85:6042–6046

Schaeffer WI (1980) The long-term culture of a diploid rat hepatocyte cell strain. Ann NY Acad Sci 349:165–182

Sharma HW, Sokoloski JA, Perez JR, Maltese JY, Sartorelli AC, Stein CA, Nichols G, Khaled Z, Telang NT, Narayanan R (1995) Differentiation of immortal cells inhibits telomerase activity. Proc Natl Acad Sci USA 92:12343–12346

Sedivy JM (1998) Can ends justify the means?: Telomeres and the mechanisms of replicative senescence and immortalization in mammalian cells. Proc Natl Acad Sci USA 95:9078–9081

Shields R, Smith JA (1977) Cells regulate their proliferation through alterations in transition probability. J Cell Physiol 91:345–355

Smith JA, Martin L (1973) Do cells cycle? Proc Natl Acad Sci USA 70:1263–1267

Smith JR, Whitney RG (1980) Intraclonal variation in proliferative potential of human diploid fibroblasts: stochastic mechanism for cellular aging. Science 207:82–84

Todaro GJ, Green H (1963) Quantitative studies of the growth of mouse embryo cells in culture and their development into established lines. J Cell Biol 17:299–313

Tokiwa T, Nakabayashi H, Miyazaki H, Sato J (1979) Isolation and characterization of diploid clones from adult and newborn rat liver cell lines. In Vitro 15:393–400

Wilmut I, Schnieke AE, McWhir J, Kind AJ, Campbell KH (1997) Viable offspring derived from fetal and adult mammalian cells. Nature 385:810–813

SV40-Mediated Immortalization

H. L. Ozer[1]

1
Introduction

Simian virus 40 (SV40) has been extensively used as a model system for mammalian cell replication and gene expression and has served as a highly effective "probe" for cellular functions. This has also been the case for understanding carcinogenesis since viral gene expression can result in altered cell proliferation and appearance of multiple "transformed" cell phenotypes associated with tumors. Both human and rodent SV40-transformed fibroblasts show reduced growth factor (i.e. serum) requirements, multilayer growth with increased saturation density and focus formation, and anchorage independence with growth in semi-solid media such as agar or agarose. Most relevant to this chapter, SV40 also increases the frequency at which cells become immortal. It should be noted that SV40 affects a wide variety of cell types (Tooze 1981) but this discussion will focus on fibroblasts, for convenience. Its effect on immortalization is particularly striking in human cells. Mouse cells have a quite limited life span in culture (approximately 10–20 generations) but can spontaneously become immortal at a measurable frequency; infection with wild-type SV40 results in almost 100 % occurrence (Tevethia et al. 1998). On the other hand, normal human diploid fibroblasts virtually never spontaneously become immortal even within the greater life span of 50–60 generations observed with newborn or fetal cells in culture (McCormick and Maher 1988). Stable introduction of SV40 results in the appearance of the transformed phenotype in cells expressing viral T antigens, yet these cells are not immortal, although the life span is extended beyond that of human fibroblasts (Neufeld et al. 1987). Hence we will use the term immortal to refer to an additional property, rather than include it within the general complex of the transformed phenotype as commonly defined. Such SV40-transformed human cells (SV/HF) also show an increased likelihood of becoming immortal, albeit at a low frequency (between 10^{-5} to less than 10^{-8} per cell) in most cases (Neufeld et al. 1987; Shay and Wright 1989). This provides the investigator the opportunity

[1] Department of Microbiology and Molecular Genetics UMD-New Jersey Medical School 185 South Orange Avenue Newark, New Jersey 07103–2714, USA.

Progress in Molecular and Subcellular Biology, Vol. 24
A. Macieira-Coelho
© Springer-Verlag Berlin Heidelberg 1999

to develop genetically matched non-immortal SV/HF and immortal clonal and non-clonal SV/HF sublines derived from them. Such matched sets serve to identify putative growth suppressor(s) and other functions involved in immortalization. For convenience the term SV/HF will be used to denote human fibroblasts bearing and expressing an SV40 genome, and SVtsA/HF for those cells with an SV40 genome expressing a temperature-dependent large T antigen.

2
SV40 Genome and Gene Expression

The SV40 genome is a circular, double-stranded DNA of 5.2 kb. Its sequence is known and multiple mutations in virtually all of its functions are available (see Tooze 1981 or Cole 1996 for more detailed descriptions.) Its genome can be divided into three regions: a control region, the region expressed early in infection (early region), and the region expressed late in infection (late region). The control region contains the sequence required for DNA replication (origin) flanked by the promoter and the enhancer for the early region as well as the promoter for the late region. The early region is required for viral DNA synthesis. The late region encodes the viral structural proteins. The course of SV40 infection is species dependent. It replicates efficiently in several continuous monkey cell lines, especially CV-1, BSC-1 or VERO as well as primary AGMK cells. Infection of rodent cells is considered non-permissive, as it produces little or no progeny virus. Replication of its DNA is undetectable in established mouse or rat cell lines and is markedly reduced ($\sim$100-fold) in Chinese hamster cell lines (LaBella and Ozer 1985). SV40 infection of human cells is semipermissive. Low but readily detectable levels of progeny virus are produced (typically 10^6 infectious units per culture of 10^6 cells). However, viral DNA synthesis is quite efficient in many cells with as many as 10000 copies per cell reported (Ozer et al. 1981). Infection of human fibroblasts by SV40 virus results in a mixed cell population. Some but not all cells are productively infected, producing high levels of DNA and virus, and die within a few days due to the cytopathic effect of the virus. Due to spread of progeny virus, most cultures establish a persistent infection with a chronic very low level of virus production. In addition, the viral genome is integrated into many cells, resulting in their transformation. However even these cells are capable of undergoing spontaneous excision and subsequent replication of the viral genome (Girardi et al. 1965; Zouzias et al. 1980).

The SV40 early region encodes three polypeptides with overlapping sequences. These early sequences are expressed in transformed cells in contrast to those of the late region which are rarely expressed in non-permissive transformed cells and are not essential to transformation. There is a single early transcript which is processed through alternative splicing to generate three stable mRNAs which encode the large T antigen, small t antigen, and tiny T antigen. All mRNAs are capped at the 5' end and polyadenylated at the 3' end at a common sequence. All three gene products share the N-terminal amino acid sequence in frame (initial codon at nucleotide 5160 on the SV40 genome), suggesting overlapping as well as unique functions. There are four predicted α-helical structures in this region (Brodsky and Pipas 1998). Large T antigen, which is essential to viral replication and to transformation, will be used as the reference polypeptide. Its mRNA is composed of two exons with an intron between nucleotide 4917–4572. The coding region ends at nucleotide 2693 on the SV40 genome. The protein has 708 amino acids (aa; apparent molecular weight of 90 kDa on SDS-PAGE electrophoresis) and has been dissected into a series of functional domains as will be discussed in detail later. Small t antigen shares all of the first exon such that aa 1–82 are identical to large T antigen. It, however, does not use the splice donor at nucleotide 4917 and the polypeptide contains an additional 92 amino acids. It has an apparent molecular weight of 20 kDa. Since these additional sequences are within the intron of large T antigen, it is readily possible to generate deletion mutants of SV40 which selectively inactivate small t antigen either by truncation resulting in an unstable polypeptide or by failure to properly splice its mRNA. Such "dl" mutants are viable for replication in cell culture, leading initially to the conclusion that they were non-essential. Subsequently, it was appreciated that small T antigen has functions which not only affect the function of large T antigen but also ones which are independent of large T antigen in transformation, particularly in studies involving cells which were not proliferating at the time of infection with SV40. Tiny T antigen shares the first 131 amino acids with large T antigen involving exon 1 and shares the splice acceptor and the first 49 amino acids of exon 2 but uses a cryptic splice donor and acceptor site to generate four unique amino acids at nucleotide 4425 (ALLT) at its C-terminus, with an apparent molecular weight of 17 kDa. It is present at low levels and few studies on its role have been reported (Zerrahn et al. 1993). A wide variety of mutations are know for large T antigen, which represents the A genetic complementation group of the virus genome. A database of mutations is accessible through a website (http://www.pitt.ed/-pipas lab; Robinson and Pipas 1998). These include mis-sense mutations which encode a heat-labile temperature-sensitive large T antigen polypeptide (termed tsA mutants), all of which map in the unique region. Although mutants in the other T antigens are also available (e.g. uniquely for small t antigen and through common sequences for all T antigens), conditional ts mutants are not presently available for small t or tiny T antigens.

Prior to assessing the role of SV40 in immortalization, I would like to address its diverse functions. First, this virus needs to be considered in the context of its replication requirements in its permissive host. SV40 infects monkeys in nature and replicates most efficiently in primary or continuous cell lines of monkey kidney cells. Second, replication of its genome is very dependent on cell functions. Although large T antigen is essential to viral

DNA synthesis, having both initiation and helicase functions, all other proteins involved in viral DNA synthesis are cell-encoded. Consequently, the virus is dependent on the level of these proteins, which are typically highest during S phase of the cell cycle. Third, SV40 infects cells (e.g. renal epithelial cells) which typically do not proliferate *in vivo*. Hence, it might be expected to have evolved functions which promote the induction of cell progression into S phase. Finally, SV40 does not encode any polypeptide which directly causes transcription (e.g. RNA polymerase) or RNA processing. Therefore it is profoundly dependent on host functions in multiple ways.

Befitting this multifaceted dependence on cell functions, it is not unexpected that large T antigen is widely distributed within the cell and impacts directly or indirectly on a wide range of cell processes. It is preferentially localized in the nucleus, and can bind to DNA (Gruss et al. 1988) but a minor percentage of large T antigen also has been identified as an integral plasma membrane protein (Deppert 1980). Most importantly it has been found associated with numerous cell polypeptides, presumably altering their functions. These associations have been most often documented by the isolation of intracellular complexes, *in vivo* binding studies, and mutation analyses. Since most of these "target" polypeptides are involved in cell proliferation, at a minimum they contribute to viral-mediated transformation and immortalization.

Large T antigen function can be dissected into specific domains. Initial studies took advantage of the feature that deletion mutants were still compatible with a stable polypeptide; subsequent analysis identified mis-sense mutations in many cases. It should be emphasized, however, that these analyses have not invariably yielded concordant results, as would be expected if such functions involve interactions in different regions of the polypeptide. In addition, since immortalization and transformation are studied in both human cells (i.e. a primate related to its natural host) and rodent cells, it is possible that the interaction of the T antigens with the respective cell proteins may not be identical. This premise is most readily demonstrated by analysis of the host range of SV40 itself since rodent cells do not support viral DNA synthesis; detailed biochemical analysis has shown that it is due to the inefficiency of the interaction of large T antigen with the DNA polymerase α holoenzyme, resulting in the failure to initiate viral DNA synthesis in such cells or in purified *in vitro* preparations (Stadlbauer et al. 1996). With these caveats, large T antigen has been shown to have the following domains. It has a nuclear localization signal (aa 126–132) and an adjacent DNA binding domain (aa131–259). It binds to cellular chromatin; however, the role is not clearly understood (Hinzpeter and Deppert 1987). On the other hand, its ability to bind to viral DNA is sequence-specific and essential to the initiation of viral DNA synthesis at a discrete origin (site II). Large T antigen also binds at site I in the control region between site II and the start of the early transcript, serving to autoregulate (downregulate) its own synthesis. A DNA helicase activity has been mapped broadly to aa 131–627.

Binding sites for DNA polymerase (pol α) have been identified in the N-terminal aa 1–82 and elsewhere. All of these functions have clear roles in viral replication; however, their roles in transformation are not readily evident, since SV40 does not replicate in the established mouse cell line 3T3, a popular assay system for transformation. Indeed, large T antigen weakly binds subunits of the mouse pol α complex as noted above. Nonetheless, the possibility of direct effects on initiation of cellular DNA synthesis needs to be borne in mind.

It is clear, however, that SV40 large T antigen has a profound role in cellular gene expression as well as on gene function. First, it affects a number of general transcription factors which act on promoters involving RNA polymerase II (pol II), typically increasing their level of expression (Saffer et al. 1990) and/or binding to them directly (Damania and Alwine 1996). Since these interactions are not limited to the permissive cell, they have a bearing on transformation and/or immortalization as well. In addition, large T antigen co-activates ribosomal RNA transcription which involves pol I (Zhai et al. 1997). In many cases the specific sites on large T antigen have not been identified; a role for the N-terminal region is frequently observed but it is likely that more than one domain may be involved (Zhu et al. 1991).

Secondly, the interaction of large T antigen with the retinoblastoma susceptibility protein (pRB) and other RB-family members (p130, p107) plays an essential role in virus replication and transformation (Weinberg 1995; Tara 1997). The sequence in large T antigen has been localized to aa 101–118 and requires the amino acids LxCxE at 103, 105 and 107 respectively; a missense mutation at aa 107, which substitutes lysine for glutamine (termed the K1 mutant) results in loss of binding for all three polypeptides (DeCaprio et al. 1988). Since these RB proteins bind the E2F family of transcription factors as well as bind to DNA directly, the role of large T antigen is mediated through its effect on transcription. The role in transformation will be discussed more fully in a subsequent section. The LxCxE motif is shared with other RB-binding proteins including other DNA tumor viruses; i.e. mouse polyomavirus large T antigen, adenovirus E1A, and human papilloma virus (HPV) E7. It should also be noted that there are multiple serines and threonines in this region of large T antigen which have been shown to be differentially phosphorylated, specifically S106, S111, S112, S120, S123, and T124. The combination of phosphorylation of T124 and the lack of phosphorylation of S120 and S123 are critical to the ability of large T antigen to initiate viral DNA synthesis *in vitro*; it has been shown that the cellular phosphatase PP2A can mediate the selective dephosphorylation of S120 and S123 *in vitro* (Scheidtmann et al. 1991b).

A third aspect is provided by the recent identification of a domain at aa 42–47 (HPDKGG) which includes an invariant HPD sequence at aa 42–44. In contrast to the RB-binding domain, this is present in small t antigen as well. It has been termed the J domain because of its homology to the DNA J region in *E. coli*. Indeed, the SV40 sequence will substitute for the corresponding sequence in *E. coli* (Kelley and Georgopoulos 1997). In bacteria, the DNA J region functions as a chaperone, mediated in part by the interaction with the DNA K product which has an ATPase activity. Consistent with this model, large T antigen binds specifically to a member of the Hsp70 family of heat shock proteins (Sawai and Butel 1989) and mutation in the HPD residues disrupts binding to Hsc70 (Campbell et al. 1997). Chaperone function might be expected to impact on conformation of multiprotein complexes and can therefore be expected to affect diverse cellular processes. A model in the context of viral replication is provided by the fact that the bacteriophage λ uses a J function in the initiation of viral DNA replication by remodeling multiprotein complexes at the origin (Alfano and McMacken 1989); this domain in SV40 promotes efficient viral DNA replication (Campbell et al. 1997). The domain affects transformation as well (Porras et al. 1996; Stubdal et al. 1997).

A fourth major function of large T antigen in transcription involves the binding of the cellular growth suppressor p53 (Ko and Prives 1996; Giaccia and Kastan 1998). The domain extends over a number of amino acids in the C-terminus, involving aa 351–626. This C-terminal region also includes a zinc finger domain (aa 302–320), a putative polα binding site, and an ATPase activity (aa 418–627) associated with the DNA helicase function. Selective deletion mutations in this region have led to the identification of a bipartite domain for p53-binding encompassing aa 351–450 and 533–626; single mis-sense mutations are known which markedly affect binding and function of p53 (Kierstead and Tevethia 1993). One example is a substitution of glutamic acid for aspartic acid at amino acid 402 (D402E). Interestingly, it markedly interferes with binding of human p53 but has only a moderate effect on mouse p53 (Lin and Simmons 1991). T antigen binds p53 in the central region of the sequence-specific DNA binding domain of p53, blocking its transcriptional activation function and DNA binding activity (Bargonetti et al. 1992). It is less clear whether complex formation blocks p53's gene repression function in the N-terminus or its non-specific DNA-binding domain in the C-terminus. Large T antigen does interfere with the ability of p53 to induce growth arrest and induce apoptosis, and therefore in its ability to function as a tumor suppressor (Ko and Prives 1996). Furthermore, other DNA viruses such as adenoviruses and HPV also interact with p53 and those interactions are essential to their respective transforming activity; however, they do not bind the same region of p53 as does large T antigen. Adenovirus E1B-encoded p55 binds to the N-terminus of p53 which includes its transcriptional activation domain but does not involve its DNA binding domain (Yew et al. 1994). The E6 gene of tumorigenic HPV16 binds p53 together with the E6AP (an E3 ubiquitin ligase) and promotes the efficient ubiquitination of wild type p53 which results in its very rapid turnover (Scheffner et al. 1990). Wild type p53 is normally degraded by the ubiquitin pathway (Chowdary et al. 1994) with a half-life of only 20–30 min, but E6 results in virtually undetectible levels of p53 intracellularly and would therefore be expected to interfere with all functions of p53. Interestingly, neither the mouse polyomavirus nor the bovine papillomavirus (BPV-1) encode a polypeptide which binds p53, suggesting that the role of wild type p53 in interfering with transformation can be provided by other gene products and might be a subset of the multiple functions by which p53 affects the cell.

A fifth aspect involves the function of a major transcriptional coactivator, the histone acetyltransferase p300 (Shikama et al. 1997) although the mechanism by which large T antigen acts is somewhat controversial. Since p300 is bound by p53 (Lill et al. 1997a), large T antigen binds p53, and sequences in the C-terminus of large T antigen are involved in the association with p300 (Lill et al. 1997b), it has been postulated that the co-precipitation of large T antigen and p300 from cell extracts is, in fact, mediated by their common binding of p53. If so, it might restrict the effects of p300 to those involving p53-mediated effects. Other data, however, indicate that large T antigen may directly affect p300 function. One line of genetic evidence is based on the fact that E1A (which does not bind p53) binds p300, and large T antigen will complement an adenovirus mutant in E1A which cannot bind p300. Furthermore, the N-terminal domain of large T antigen is involved and mutations in aa 19–28 interfere with this ability (Yaciuk et al. 1991). A plausible explanation for the discrepant results would be that more than one region of large T antigen is involved.

Small t antigen also contributes functions relevant to cell proliferation. It shares the first 82 amino acids with large T antigen and may therefore carry out the same functions in these domains. Small t antigen can stimulate the ATPase of Hsc70, consistent with its having a functioning J domain (Srinivasan et al. 1997). It has also been reported that small t antigen is capable of transactivating RNA pol II and pol III-directed transcription from promoters on reporter plasmids (Loeken et al. 1988) and, in a more limited number of cases, on an endogenous gene, including the pol II-dependent promoter of cyclin A (Porras et al. 1996). This function also requires aa 42–47. In addition, small t antigen, in contrast to large T antigen, can activate signal transduction pathway(s), including the MAP kinase pathway and cyclin D1, the regulatory subunit of the cyclin kinase D (Watanabe et al. 1996). The unique region of small t antigen (aa 83–174) contains a cysteine rich region involving a zinc-binding domain of two sets of CxCxxC (aa 111–118 and 136–143) and a unique CxxxPxC sequence at aa 97–103. This total region is involved in binding of small t antigen with the serine/threonine phosphatase PP2A (Mungre et al. 1994). PP2A is composed of three subunits (ABC); small t antigen can substitute for the B subunit, resulting in AtC enzyme with reduced PP2A activity (Yang et al. 1991). Mutational analysis has shown that the ability of small t antigen to activate cyclin D1 is reduced in constructs which have single mis-sense mutations in aa 97–103. Since only a limited number of substrates of PP2A have been identified, the full range of effects is as yet undetermined.

3
SV40-Mediated Transformation

SV40 induces a transformed phenotype in diverse species. Major aspects of these effects are mediated by the essential expression of large T antigen. Mutants of multiple large T antigen domains have shown it to be required for stable expression of transformation. Furthermore, persistent expression of large T antigen is required, as demonstrated by studies on transformation of immortal rodent cells with a conditional temperature-dependent large T antigen (e.g. SVtsA58) which show the transformed phenotype at 33° C but revert to a non-transformed phenotype when such cultures are shifted to 39° C (Brockman 1978). Since these mutations map to the unique region of large T antigen, they do not result in decreased small t antigen protein levels. Rather, they may result in its increased mRNA level due to failure of large T antigen to inhibit early region transcription by reduced binding of the tsA protein to site I on the viral genome. Persistent viral gene expression is maintained through an integrated viral genome. Expression of large T functions in both the N-terminus and C-terminus regions of large T antigen, and interactions with known tumor/growth suppressors are key features, although results differ somewhat with the experimental system employed (e.g. Maclean et al. 1994). Interaction of large T antigen with p53 is involved as noted earlier. Major effects also involve the interaction with pRB since the K1 mutant is defective in inducing transformation. Although binding is mediated solely by the LxCxE domain, functional effects require the involvement of the J domain as well (Porras et al. 1996; Zalvide et al. 1998). In the case of the RB-family members p130 and p107, association with large T antigen having both domains intact results in alteration of p130 and p107 phosphorylation (Stubdal et al. 1996), and in targeting p130 for subsequent ubiquitination and degradation. Studies with mouse embryo fibroblasts lacking pRB have suggested that these proteins may be important targets in transformation as well, even through their inactivation is not associated with human cancer in contrast to pRB (Stubdal et al. 1997). Mutations in the J domain interfere with transformation as assayed by growth to high cell density and at low serum concentration, but is not evidently required for anchorage independence.

Not only are these and other large T antigen-dependent functions involved in transformation but small t antigen functions are involved as well. Introduction of large T antigen in the absence of small t antigen does not result in the full transformed phenotype. In the case of rodent cells (e.g. rat F111 fibroblasts) there is a reduction in the frequency of anchorage–independent colonies; mis-sense mutations in the PP2A unique domain (aa 97–103) generate a similarly defective phenotype (Mungre et al. 1994). In the case of normal diploid human fibroblasts, there is a defect in focus formation (deRonde et al. 1989). This defect is similarly observed with mutants in the PP2A domain; however, it is also seen with mutations in other regions

as well (Porras et al. 1996). For example, SV40 mutant 5002 (L19F/P28S substitution) or changes in the J domain do not generate foci when the mutation involves both large T antigen and small t antigen. Co-transfection with
plasmid encoding each mutant large T antigen can be complemented in
trans by wild type small t antigen. These results cannot simply be explained
by effects on PP2A. Furthermore, although in general the requirement for
small t antigen can be best demonstrated when quiescent cells are employed,
since it has been repeatedly shown that small t antigen facilitates G1 progression (e.g. Cicala et al. 1994), the results described above are not
explained solely by such a mechanism since stimulation by serum growth
factors can compensate for the effects on the MAP kinase pathway but do
not provide the functions in the mutant 5002 or J domain. Small t antigen,
like large T antigen, also transactivates the cyclin A promoter (Porras et al.
1996). It must be further considered that small t antigen acts through multiple mechanisms involving large T antigen , not only those shared due to the
overlapping aa 1–82 but also because of the fact that large T antigen itself
may be a target of PP2A. Other targets of PP2A based on *in vitro* studies
include p53 (Scheidtmann et al. 1991a).

4
SV40-Mediated Immortalization

4.1
Extension of Life Span

With the preceding information as background, one is better able to assess
the role of SV40 in immortalization of mammalian cells. Normal cells have
a limited life span in culture (Hayflick 1965), undergoing replicative senescence. Long considered a model for aging at the cellular level (i.e. the "Hayflick limit"), it has become recognized as a potential tumor suppression
mechanism as well (Campisi 1996). Replicative senescence is characterized
by the irreversible arrest of cells in the G1 state of the cell cycle (Goldstein
1990), with a pattern of gene expression altered from that of replicating cells
and cell morphological changes. Such cells remain viable in culture and presumably *in vivo*, providing a mechanism whereby they may also affect the
behavior of other non-senescent cells, either directly by cell-cell contact or
through their secreted products. Replicative senescence can be viewed as
one of the limited number of responses of individual mammalian cells to
adverse conditions. For example, normal fibroblasts can arrest in G_0, a
reversible balanced non-growth state, when presented with insufficient
growth factor. Alternatively, cells may undergo programmed cell death or
apoptosis, including, but not limited to, conditions of deranged intracellular
gene expression or in response to extracellular stimuli. In this context replicative senescence provides a third, possibly intermediate, option. Like G_0-
arrest, the cells remain viable and can express gene functions compatible

with the differentiated phenotype of that cell type – a possibly beneficial property for the intact organism. On the other hand, it shares with apoptotic cells the feature of being irreversible. This latter feature has particular relevance to replicative senescence as a potential tumor suppression mechanism. It has been shown that overexpression of an activated oncogene (i.e. ras) by DNA-mediated gene transfer can acutely induce a replicative senescent phenotype in human diploid fibroblasts (Serrano et al. 1997). Thus, one might speculate that deregulated expression of ras due to mutation could trigger an intrinsic senescence response in a cell, resulting in the subsequent failure of that cell to proliferate within the organism. Alternatively, the programmed onset of replicative senescence would result in the irreversible growth inhibition of previously mutated cells. In either case, the putative tumor cell fails to proliferate and therefore does not accumulate the multiple alterations requisite for carcinogenesis. In this context, replicative senescence is therefore a fundamental response of the cell to intrinsic (or extrinsic) insult. It serves as an alternative to apoptosis, in which the cell is not merely arrested but dies (and disappears). In both mechanisms, circumvention is predicted to be carcinogenic. If replicative senescence functions as a tumor suppression mechanism, tumor cells should have bypassed or overcome senescence. Indeed, human tumors when put into cell culture commonly grow continuously and are termed immortal (although it is somewhat controversial whether this phenotype is present *in vivo*; Stamps et al. 1992). Since both human and mouse cells have been commonly used as experimental systems, it is often possible for one to draw shared insights from both systems. It must be stressed, however, that significant differences exist between cells from the two species, especially in the propensity to undergo immortalization spontaneously and in the presence of SV40 gene expression as well. In addition, it is useful to bear in mind that SV40 infection of human cells, in contrast to rodent cells, is semi-permissive. To minimize the complexity of the virus-human cell interaction, several laboratories have used approaches such that replication is precluded. We have emphasized the use of SV40 genomes that have a 6 base pair deletion at the BglI site in the origin of DNA replication (Small et al. 1982). Such origin-defective mutants cannot replicate the viral genome, eliminating complications due to viral cytopathic effects (Gluzman et al. 1980). Furthermore, when such sequences are introduced into human cells by DNA-mediated gene transfer only one or a few copies of SV40 sequences become integrated and are persistently expressed. The absence of free unintergrated viral sequences also facilitates assessment of functional and structural viral changes (Neufeld et al. 1987).

SV40 gene expression, although necessary is not, however, sufficient for immortalization, resulting in the proposal of a two-stage model by us and others. Wright et al. (1989) emphasize an M1 restriction associated with replicative senescence, and a subsequent restriction termed M2 which also needs to be passed to become immortal. We (Radna et al. 1989; Ozer et al. 1996) define a stage I after introduction of SV40 sequences which results in the bypass of senescence and the extension of life span. This effect is, however, of limited duration with the cells subsequently undergoing "crisis" with cell death. Stage II involves those additional changes which result in continuous cell proliferation beyond crisis, i.e. immortalization. Many of the changes in stage I are also involved in the eventual immortal phenotype of the cell. Effects of SV40 in stage I parallel the effects on transformation; however, large T antigen alone can extend the life span of fibroblasts and promote immortalization. When fibroblasts are transfected with SVtsA sequences they show extended life span at 32–35° C. However, when such cultures of either human (Hubbard-Smith et al. 1992) or mouse (Ikram et al. 1994) cells are shifted to 39° C and large T antigen function becomes inactivated, cells show altered ability to continue to proliferate. The cells resume the stage of life span of the uninfected cell, i.e. they behave as if they had undergone an equivalent number of generations to that which they had previously at 35° C. If the cell is within the "normal" life span, the cells continue to proliferate for a few or several generations before arresting in a senescence-like phenotype. If they are beyond the normal life span, they undergo growth arrest and/or cell death within one or a few generations. For human fibroblasts, for example, this occurs even in immortal SVtsA/HF at 39° C (Radna et al. 1989) or SV/HF in which large T antigen expression is hormone dependent (Wright et al. 1989), demonstrating the reversible nature of the large T antigen effect. In addition, these data show that there is a requirement for functional large T antigen for cell proliferation even in cells which have undergone SV40-mediated immortalization.

The role of large T antigen in extension of life span and immortalization is dependent on its ability to interact with the growth suppressors pRB and p53 since mutants of large T antigen which do not express either of these functions are defective in extending life span. Similarly, non-immortal and immortal cells generated with tsA mutants of large T antigen lose both binding properties upon temperature shift from 35 to 39° C (Resnick-Silverman et al. 1991; Hubbard-Smith et al. 1992). pRB plays a critical role in replicative senescence, and the ability of large T antigen to bind and inactivate pRB is central to the ability of large T antigen to overcome senescence and promote immortalization. RB family members bind E2F family members. E2F members form complexes with DP-1 or DP-2 and bind to DNA in a sequence-specific manner. They function as activators of cellular genes, including ones involved in DNA synthesis and S phase entry (Slansky et al. 1993). The pRB-E2F complex also binds histone deacetylase (HDAC1) whose action affects chromatin structure and inhibits the action of E2F as a transcription factor (Brehm et al. 1998; Magnaghi-Jaulin et al. 1998). pRB binds E2F-1, E2F-2, E2F-3 and E2F-4; p107 and p130 can bind E2F-4 and E2F-5. pRB binds to DNA itself and may similarly repress other promoters. Phosphorylation of pRB occurs by the sequential action of cyclin kinases (Tara 1997), composed of a cyclin-dependent kinase (cdk) and a regulatory subunit (cdc). Action of cyclin kinase D (cdk4:cycD1and cdk6:cycD1) and cyclin kinase E (cdk2:cycE) during late G1 results in dissociation of the HDAC1, pRB, E2F complex, facilitating gene expression (DePinho 1998). Senescent cells accumulate hypophosphorylated pRB to which E2F remains bound and is considered a pivotal restriction point in G1 (Stein et al. 1990). Large T antigen through the LxCxE motif binds pRB and in concert with the J domain mimics phosphorylation in that it results in the dissociation of E2F from pRB. In addition it may directly complete with the binding of HDAC1 with pRB, since both large T antigen and HDAC1 interact with a common site on pRB (Magnaghi-Jaulin et al. 1998). Hence, the consequences of pRB in replicative senescence are prevented. The effects of large T antigen on p107 and p130 (and E2F-4 and E2F-5) in this context are not well understood. It should be noted that p130 is preferentially expressed in G_o (i.e. at a stage different from replicative senescence).

Comparative studies with SV40 in human fibroblasts have generally involved the K1 mutant. A detailed description of the involvement of sequences at both the N-terminus and C-terminus of large T antigen has recently been reported in mouse embryo fibroblasts (MEF; Tevethia et al. 1998). When the sequence encoding the N-terminus (T1–147) is introduced into primary or secondary MEF, life span is extended beyond the four passages of normal MEF as evidenced by their ability to form colonies. However, such colonies do not progress to generating immortal sublines at high frequency. Introduction of the carboxyl terminal region (aa 251–708) gives the same result. As noted earlier, this fragment contains the bipartite p53-binding domain (Kierstead and Tevethia 1993). p53 has growth inhibitory effects on cells, which are inducible at the transcriptional and post-transcriptional level by DNA damage. Senescent cells show some elevation of p53 levels (Atadja et al. 1995) and high levels of the cyclin-dependent kinase inhibitor (CKI) p21 (Noda et al. 1994), a known target of p53-transcriptional activation function (El-Deiry et al. 1993). The effects of p21 are multifaceted but it is an inhibitor of phosphorylation of pRB (Harper et al. 1993), and human fibroblasts in which both copies of p21 are disrupted show extended life span (Brown et al. 1997). Hence, p53 contributes to the underphosphorylation of pRB in senescence. As noted earlier, large T antigen blocks the transactivation function of p53. SV40 virus encoding a mutant large T antigen (e.g. D402E), defective in binding of p53, does not extend the life span of human fibroblasts. Other factors may be relevant as well. Although senescent cells do not show enhanced apoptosis, the elevated p53 may also mediate a pro-apoptotic effect. For example, p53 induces the pro-apoptotic bax through its transactivation function (Miyashita and Reed 1995) and bax can interfere with the anti-apoptotic function of bcl2 by complexing with it (White 1996). The intracellular consequences could be important in cells expressing T1–147, for example. Elevated (free) E2F-1 has been reported to induce apoptosis (Qin et al. 1994) and the interference of T1–147 with pRB binding of E2F may serve to generate such a situation. It should also be noted that large T antigen has a putative anti-apoptotic domain in its C-terminus region (aa 525–541; Conzen et al. 1997). Hence both N-terminal and C-terminal domains of large T antigen have the ability to transiently bypass senescence but each does not provide in itself sufficient alteration for the cell to be immortal. These findings with MEF differ from earlier reports from that laboratory and others, and are likely due to the very limited life span of these MEF in the absence of SV40 sequences, minimizing the likelihood of other changes in MEF. Co-transfection of both constructs generated colonies which showed a 100 % incidence of generating immortal sublines, as does wild type SV40. These results are consistent with those for human fibroblasts as evidenced by the studies with SV40 mutants (K1, D402E) cited above, and experiments in which antisense constructs to mRNA encoding pRB and p53, respectively, result in co-operative extension of life span and transient bypass of senescence (Hara et al. 1991). However, in marked contrast to the results on immortalization of MEF by such an effector, stably transformed human fibroblasts containing an integrated SV40 genome encoding large T antigen (and small t antigen) do not become immortal at high frequency.

It should be noted that other effectors on which large T antigen (and small t antigen) would expect to have impact are altered in senescent cells and might therefore have bearing on SV40-mediated extension of life span. p16, a CKI of a family different from p21, is also elevated in senescent cells (Alcorta et al. 1996). A major effect is through its inhibition of (cycD-mediated) phosphorylation of pRB; hence large T antigen expressing cells should be less sensitive to inhibition by this mechanism. Indeed, SV/HF show persistent elevated levels of p16 during extended life span (Hara et al. 1996). Senescent cells have also been reported to have reduced activity of the MAP kinase pathway (Tresini and Cristofalo 1998); small t antigen stimulates MAP kinase (ERK1) phosphorylation and activity through inhibition of PP2A. Similarly, the small t antigen (and large T antigen) mediate induction of cyclin kinase activity through action on the cyclin A promoter and should act to reverse the low levels of activity of cyclin kinases described for senescent human fibroblasts (Stein et al. 1991). Marked increases in cyclin A, cyclin B and $p34^{cdc2}$ polypeptide levels are observed in SV/HF (Rubelj and Pereira-Smith 1994; Chang et al. 1997). Since there is only a modest change in their respective mRNA levels when compared with normal (non-senescent) human fibroblasts, an increased translational efficiency or protein stability has been suggested to be involved as well. The dysregulation of cyclin kinase D1 and E in senescent human fibroblasts is more complex since it includes the accumulation of inactive cyclin–cdk complexes (Dulic et al. 1993). Although small t antigen has been shown to increase the level of cyclin D1 through its effect on the MAP kinase pathway, large T antigen is purported to reduce levels of cyclin D1. In SV/HF, Peterson et al. (1995) reported quarternary complexes of cdk4 with cycD1, PCNA and p21 being disrupted while complexes involving cycD1, PCNA and p21 were still present, and levels of D1 were not reduced. Cyclin E was increased, as was cyclin A.

In conclusion, SV40 T antigens have major effects on the manifold aspects of cell cycle regulation which are altered in replicative senescence. The ability of SV40 to extend life span and block senescence requires such T-dependent functions. Human and mouse cells which are non-immortal at the time of introduction of SV40 sequences in culture remain dependent on these functions even after they become immortal cell lines. This finding is in contrast to results in which SV40 is introduced into already immortal cell lines (e.g. 3T3) in which cell proliferation is not T-dependent under standard culture conditions, although those growth conditions involving expression of the transformed phenotype typically can be shown to be T-dependent.

4.2
Crisis

This extension of life span by large T antigen (or large T antigen plus small T antigen) is, however, limited. This phenomenon is quite evident in human cells. After a continuous period of vigorous cell proliferation beyond the normal life span, cultures of large T antigen-positive cell populations show progressive slowing and the great majority of the cells die after 20–30 additional generations with the culture terminating in so-called crisis (Girardi et al. 1965; Neufeld et al. 1987). Since this is also observed in SV/HF containing an origin-defective SV40 genome, it cannot be explained by the effects of virus replication. A complex series of poorly understood events are occurring. Some cells show the appearance of aspects of the senescence phenotype, including expression of a senescence-associated $ galactosidase (Dimri et al. 1995) and overexpression of genes such as type I collagenase (West et al. 1989). Since many of the senescence-associated genes have a complex regulation and their role in replicative senescence is not understood, it has been difficult to interpret the basis for the alterations. Modest elevation of CKI p21 and persistent elevation of p16 are also evident but these inhibitors are expected to have more limited effects in the presence of large T antigen than in normal cells. Furthermore, the majority of cells die, in contrast to the behavior of normal senescent fibroblasts which undergo irreversible growth arrest but do not progressively detach from the monolayer or die. Unpublished studies from this laboratory demonstrate the progressive appearance of DNA nucleosome ladders, double stranded DNA damage by TUNEL assay, and morphological changes characteristic of apoptosis (White 1996) in the terminal subculture of SV/HF cells undergoing crisis (Lenahan 1998).

It is not clear what is responsible for crisis. It could be related to large T antigen function itself. For example, the proliferative signal of large T antigen may establish a "mixed signal" in the context of cell arrest due to senescence or another mechanism. Such dysregulated growth has been invoked to explain other situations which result in apoptosis (White 1996). Furthermore, acute expression of large T antigen induces DNA synthesis but not mitosis when virus infects synchronized permissive CV-1 cells; inviable cells accumulate in G2 (Pages et al 1973; Gershey 1979). Large T antigen similarly does not induce mitosis when introduced into senescent human fibroblasts (Gorman and Cristofalo 1985). In addition, a recent study observed that large T antigen can promote apoptosis when an SVtsA construct is introduced into already immortal (HPV-containing) cervical carcinoma cells at the non-restrictive temperature of 35° C but not at 39° C (Chen et al. 1998). The apoptotic phenotype can involve as many as 20 % of the cells on a chronic basis, and is associated with overexpression of the transcription factor c-jun. Finally, we have found that large T antigen does not block some forms of apoptosis (Lenahan and Ozer 1996). The last two studies may involve possible p53–independent mechanisms.

A second model is that a large T antigen-inducible survival factor pathway (e.g IGF-1 or IGF-1 receptor) may no longer be effective (Ferber et al. 1993; O'Connor et al. 1997). Non-immortal SV/HF do require greater serum concentration during the later period of extended life span (Stein 1985). It might be necessary to postulate a change in large T antigen function or a change in the cell death program to cause the onset of crisis in non-immortal SV/HF after several generations of vigorous cell proliferation. Alternatively, an apoptotic pathway which large T antigen has been blocking may no longer be as effectively blocked. Although the ability of large T antigen to bind p53 and therefore block p53-dependent pathways (Symonds et al. 1994) is well documented, a marked elevation of wild type p53 has been reported to cause apoptosis in SV/HF despite persistent large T antigen levels (Caelles et al. 1994). Furthermore, DNA damage can induce elevated levels of p53 in SV40 transformed mouse cells (Hess and Brandner 1997). Large T antigen has also been recently reported to have an additional non-p53, anti-apoptotic motif (Conzen et al. 1997). Under this second model, one might speculate that immortal SV/HF, which have bypassed or survived crisis would have an altered response to apoptotic inducing agents. They might also show altered expression of genes which affect the function of such an effector. Mutations in p53 have not been observed in immortal SV/HF (Moorwood et al. 1996). On the other hand, we have recently described a set of genes which are underexpressed in several immortal SV/HF, including a serine/threonine PP2C-phosphatase which has been reported to interact with components of signal transduction pathways (Pardinas et al. 1997; Takekawa et al. 1998).

A third model is that the length of chromosomal telomeric sequences may represent the mechanism for such a large T antigen-independent apoptotic and/or senescent signal in SV/HF. It is well-accepted that normal human fibroblasts do not express the catalytic subunit of telomerase (hTERT), an essential gene in the regulation of telomere length (Harley et al. 1990). Consequently, human fibroblasts, like other somatic cells, show progressive shortening of their telomeres as evidenced by analysis of the size of a terminal restriction fragment (TRF), for example. Although the mechanism is unknown, it has been postulated that replicative senescence is triggered by a critical level of TRF (Allsopp et al. 1992). There are some uncertainties in this hypothesis since both human and mouse fibroblasts undergo replicative senescence, yet the latter still have very long TRF ($>$ 40 kb; Sedivy 1998). Nonetheless, this hypothesis has received strong support from the recent report that introduction of the cDNA for hTERT can restore telomerase activity and generate TRF lengthening, indicating that it, rather than other components, e.g. telomere-specific RNA (Blasco et al. 1997) and/or telomere binding proteins (van Steensel and de Lange 1997), is the key defect in such cells (Bodnar et al. 1998). Moreover, life span is markedly extended. SV40 does not reactivate telomerase; hence TRF shortening proceeds further during the extended life span of SV/HF (Counter et al. 1992; Small et al. 1996). If replicative senescence is mediated by a mechanism which large T antigen can obviate (e.g. a p53-dependent checkpoint), SV/HF could bypass the growth arrest observed in normal cells. Crisis could then result from the inability of large T antigen to continue in that role due to the involvement of other factors which are not responsive to large T antigen, or the eventual overwhelming of the ability of large T antigen to inhibit the primary mechanism (e.g. due to excessive levels of p53). Alternatively, the further shortened TRF might generate new phenotypes which themselves promote cell death. Such a mechanism might include end-to-end joining of chromosomes, which is known to occur in the absence of adequate telomeres. Large T antigen would not be expected to be able to promote survival of such cells.

4.3
Isolation of an Immortal Cell Line

A rare SV/HF cell becomes immortal. Since all SV/HF cells express large T antigen and there is no evidence for major changes in SV40 function in the immortal, compared with the pre-immortal SV/HF (Hubbard-Smith et al. 1992), attention has focused on non-SV40 functions being responsible. Several lines of evidence support this interpretation. First, immortalization is recessive to limited life span in cell hybrids (Pereira-Smith and Smith 1981). Hybrids between senescent cells and normal cells have a reduced life span consistent with the presence of an inhibitor of cell proliferation in senescent cells. Large T antigen would be expected to show a dominant effect and override senescence. Since hybrids between immortal and normal cells have a limited life span, it would appear that immortal cells are not resistant to such inhibition (Pereira-Smith and Smith 1983). Thus, the data support the model that immortal cells have inactivated their own senescence (SEN) gene. Second, hybrids between immortal cells are often not immortal. Pereira-Smith and Smith (1988) have shown that hybrids between independent immortal cell lines behave in a fashion consistent with complementation of loss of function. In the case of immortal SV/HF, all but one of the SV/HF that they tested did not complement each other; multiple cell lines did complement SV/HF and each other. Overall the data fit a minimum of four complementation groups (A-D), with many independent SV/HF and several human tumor cell lines in group A.

An additional role for p53 might be considered as a cellular function responsible for immortalization. First, it has a clear role in extension of life span in SV/HF. Secondly, in some (but not all) mouse strains, p53 mutations are a major factor in establishing immortal sublines from normal animals (Harvey and Levine 1991). Thirdly, MEF from p53 knockout mice immortalize readily (Harvey et al. 1993). Human fibroblasts from patients with Li-Fraumeni syndrome bearing missense mutation(s) in p53 show an increased frequency of "spontaneous" immortalization when compared with normal cells (whose frequency is nil; Bischoff et al. 1990). p53 has a complex pattern of function and regulation in general, and in SV/HF in particular. Large T antigen not only forms complexes with p53 and inactivates function(s), it also stabilizes wild type p53, suggesting that it alters p53 function in other ways as well. Stabilization is thought to be mediated through binding to large T antigen but that may not be the only mechanism (Deppert et al. 1987). Small t antigen has also been reported to stabilize p53 even though it does not bind to it (Tiemann et al. 1995). It is not known whether this is mediated (solely) by alteration of the effects of PP2A on p53 phosphorylation; specific sites of phosphorylation provide a regulatory mechanism for p53 function (Giaccia and Kastan 1998). The stability of p53 is also an important consideration in the interpretation of most studies involving non-immortal and immortal SV/HF since they have generally focused on the level of (immunoreactive) p53. p53 is indeed complexed by large T antigen but significant levels are not in both types of SV/HF (Lenahan 1998). O'Neill et al. (1997) even reported that most p53 is not in complexes with large T antigen in some immortal SV/HF; this result is due to an excess of p53 over large T antigen levels in those cells, consistent with prior reports that large T antigen binds only a subfraction of wild-type p53 molecules in some cell types (Deppert et al. 1987). The function of uncomplexed stabilized wild-type p53 is unlikely to be equivalent to induced p53; indeed, many mutant forms of p53 are more stable than the wild type. However, the implications of the presence of free p53 in SV/HF must be considered. On the other hand, other data make it unlikely that p53 or its altered function are responsible for immortalization of SV40. First, the aforementioned results with Li-Fraumeni fibroblasts also showed that the frequency of immortalization is very low, not observed in all experiments, and the immortal cell lines show multiple chromosomal rearrangements indicating that other genetic changes had occurred. Furthermore, p53 appears to remain wildtype upon immortalization of SV/HF, although only limited studies have been reported (Moorwood et al. 1996). One might expect p53 mutations to be highly selected for in the event that effects mediated by p53 not bound to large T antigen were a key aspect of immortalization. Finally, the complementation studies by Pereira-Smith and Smith (1988) showed that HeLa cells were in group B. Since HeLa cells are from a patient with cervical carcinoma infected by HPV which also encodes viral proteins which bind to p53 (and pRB) and inactivate their functions, yet HeLa and SV/HF are immortal for different reasons, it argues that inactivation of p53 is insufficient for immortalization, even in the presence of large T antigen. HPV E6 can degrade free p53 and that bound to large T antigen in SV/HF (Lenahan 1998).

One might propose that the key non-viral function responsible for immortalization of SV/HF is telomerase (hTERT), based on the results of Bodnar et al. (1998). As one would expect, immortal cells have solved the telomere end replication problem, since a cell cannot divide indefinitely in the face of critically shortened or non-functional telomeres. Hence reactivation of hTERT expression may be a biomarker of the immortal cell. Interestingly, this premise supports the model that immortality is a relevant phenotype *in vivo* since many diverse types of tumors show a high proportion of telomerase-positive cells and it has been shown to be found in colon carcinoma but not benign adenomas, for example (Kim et al. 1994). Indeed, SV/HF have solved the problem as well (Counter et al. 1992; Small et al. 1996). Most immortal sublines express hTERT. However, a significant minority have not reactivated hTERT, yet have stabilized their telomeres (Bryan et al. 1995; Small et al. 1996). It is becoming evident that multiple factors are involved in the regulation of telomerase activity in addition to the level of hTERT (Blasco et al. 1997; van Steensel and de Lange 1997; Giriat et al. 1998). Furthermore, it is likely that telomerase-independent mechanisms also exist. Since SV/HF may be particularly prone to these alternate pathways, further analysis of immortal SV/HF should be particularly useful in its elucidation. Current models favor a recombinational mechanism based on studies in yeast (Lundblat and Blackburn 1993) but there is no direct support for such mechanisms being responsible in mammalian cells. It should, however, be noted that large T antigen has been reported to stimulate homologous recombination (Cheng et al. 1997); in addition, p53 has been reported to inhibit homologous recombination which can be blocked by large T antigen (Wiessmuller et al. 1996; Bertrand et al. 1997; Mekeel et al. 1997).

Two general molecular genetic approaches have been exploited to identify other genes which are involved more directly in immortalization. In one approach, mRNA populations have been examined for differential gene expression; in the second, genetic analysis has been undertaken to identify a putative growth suppressor consistent with the model of the complementation groups described above.

A variety of molecular biological approaches have been utilized to identify and isolate cDNAs which are reflective of altered stable mRNA populations in immortal cell lines. The comparison has generally been with the parental normal cell type rather than the direct pre-immortal antecedent (Schenker et al. 1994; Vellucci et al. 1995). In a recent study of this sort (Schenker and Trueb 1998), in which subtractive hybridization was used to identify cDNAs underexpressed in immortal SV/HF (Va13 cells) compared with the normal fetal lung fibroblast counterpart (WI38), 42 clones were identified which showed at least a three to ten fold reduction. The majority were previously known genes which fell into four general categories: proteins involved in the extracellular matrix (ECM), enzymes (and their inhibitors) which are secreted into the extracellular space where they can modify ECM protein, cytoskeletal proteins, and regulatory proteins which might be expected to affect adhesion, migration and/or proliferation of cells. No cDNAs involving genes which regulate cell cycle progression directly were identified. Although such an approach is very valuable in identifying "transformation-sensitive proteins" as the authors sought, it suffers from several limitations in the context of immortalization. The most relevant consideration to this discussion is that the SV/HF were long-passaged in tissue culture, such that secondary changes post-immortalization could have occurred. Conversely, changes induced by SV40 large T antigen but independent of immortalization would also be detected. For example, the reduction in fibronectin observed in this study was also reported in non-immortal SV/HF by Rubelj and Pereira-Smith (1994). Many such examples might be expected in view of the myriad effects of large T antigen on cellular gene expression noted earlier in this chapter. It should be noted that Schenker and Trueb attempted to deal with this latter issue by also asessing human tumor-derived cell lines which did not contain SV40 sequences, including a rhabdomyosarcoma and fibrosarcoma (HT1080).

The availability of matched SV40-transformed non-immortal (i.e. pre-immortal) and immortal cell human lines in this laboratory and elsewhere offers an opportunity to assess changes which are more closely associated with the process of immortalization. Satoh et al. (1994) generated a subtracted cDNA library for sequences enriched in pre-immortal but not immortal SV40-transformed YH-1 fibroblasts; three cDNAs (vimentin, NADH dehydrogenase, and a novel sequence) were identified. Imai and Takano (1992) analyzed a cDNA library of pre-immortal SV40-transformed MRC-5 fibroblasts by hybridization with a subtracted probe and identified three cDNAs. Two (heat shock 27kDa protein and follistatin) were reduced in some but not other immortal sublines. On the other hand, a cDNA encoding type I collagenase was highly expressed in cells approaching crisis and greatly reduced in two sets of matched SV40-transformed MRC-5. Only low level expression of type I collagenase was observed when early passage pre-immortal cells were examined. They further demonstrated (Imai et al. 1994) changes in binding of transcription factors to collagenase regulatory elements in cells near crisis. An interesting difference has been reported for anti-oxidant enzymes. Dutrillaux and coworkers (Bravard et al. 1992; Hoffschir et al. 1993) observed reduced levels of the Mn-superoxide dismutase (SOD) and catalase in immortal SV/HF compared with preimmortal cells. Oxidation reactions have been postulated to play a role in aging and replicative senescence (Allen 1998). Furthermore, Yan et al. (1996) reported that introduction of a plasmid expressing exogenous Mn-SOD cDNA into SV/HF can be growth inhibitory but the cells do not become senescent.

Alterations in cyclin kinase genes have also been examined. Rubelj and Pereira-Smith (1994) reported changes in cdc2, cyclin A, cyclin B, and the CKI inhibitor p21 as well as fibronectin in precrisis (i.e. pre-immortal), crisis, and immortal SV40-transformed IMR90 fibroblasts. Although significant changes in mRNA and protein level were seen in cells in crisis, there were only slight (less than two-fold) differences when early pre-immortal and immortal cell lines were compared. As previously noted there are changes in cyclin D1 in non-immortal SV/HF. In early immortal SV/HF, Peterson et al. (1995) reported further changes including reduction in levels of cyclin D1 and its binding with both PCNA and p21, suggesting that immortal SV/HF have become independent of the cyclin D1–regulated pathway. The levels and association of cyclin A and cyclin E remained elevated and unchanged from non-immortal SV/HF. These results are consistent with the earlier findings by Xiong et al. (1993) that protein complexes containing cyclin D1, PCNA and p21 are disrupted in immortal SV/HF. Taken together, however, it is difficult to ascribe immortalization to changes in the cyclin kinases or cyclin-dependent kinase inhibitors.

In an effort to expand the number of changes specific to immortalization, we analyzed cDNA libraries for changes in multiple matched sets of pre-immortal and immortal SV/HF generated in this laboratory (Pardinas et al. 1997). A λcDNA library was prepared from a pre-immortal HF-C. We screened the library with a subtracted probe enriched for sequences present in HF-C and reduced in immortal AR5 cells. A more limited screen was also employed for sequences overexpressed in AR5 using a different strategy. Several alterations in the level of mRNAs in AR5 encoding functions relevant to signal transduction pathways were identified; however, many cDNAs encoded novel sequences. In an effort to clarify which of the altered mRNAs are most relevant to immortalization, we performed Northern analysis with RNA prepared from three matched sets of independent pre-immortal and immortal (four cell lines) SV40-transformants using eight cloned cDNAs which show reduced expression in AR5. Three of these were reduced in additional immortal cell lines as well; one of unknown function was reduced in all the immortal cell lines tested; a second was a new member of the serine/threonine phosphatase PP2C class and was reduced in two of the three matched sets; and a third of unknown function was reduced in two unrelated immortal cell lines. None of the cDNAs elevated in AR5 were consistently elevated in other immortal SV/HF. Ma et al. (1998) recently reported two cDNAs which were elevated in multiple immortal SV40-transformed human embryonic kidney cells using a differential display approach. One cDNA was novel whereas the second C1R1/CROC1 encoded a member of some newly defined ubiquitin-conjugating E2 enzyme variant proteins, which share significant sequence similarily with the ubiquitin-conjugating enzymes but lack enzyme activity.

Progress in these approaches might be expected to expand rapidly in view of developments on generating cDNA markers (such as EST) and analysis of global gene expression through DNA chips or other gene array approaches. Representations of cDNA sequences of genes of known and unknown (e.g. ESTs) function in populations of early passage non-immortal, late passage non-immortal (i.e. in crisis) and immortal SV/HF might be expected to lead to new insights. All of these approaches, however, suffer from two limitations; first, they only detect mRNA changes of sufficient magnitude. Low abundance mRNAs, modest changes in mRNA levels or changes in protein functions which are not reflected in the stable mRNA level will be overlooked. Secondly, the gene products encoded by the mRNAs need to be determined to be of functional significance to immortalization. This is perhaps the greater challenge as more differences are observed. In addition, it will be necessary to distinguish those changes which are directly involved in the immortalization phenotype (i.e. "primary effectors") from those secondary effectors which are downstream of the primary effectors or occur subsequent to immortalization.

One approach to identify such primary effectors utilizes a molecular genetics strategy. We and others have sought to identify putative growth suppressors which need to be inactivated upon immortalization by the approaches successfully utilized for isolation of tumor suppressor genes. As noted earlier, in complementation studies involving cell hybrids, SV40-transformed immortal cell lines were predominantly in the A group although subsequent studies observed a greater heterogeneity of results with SV/HF (Duncan et al. 1993). Specific human chromosomes (1,4,7) have been reported to suppress growth of immortal cells of the B-D complementation groups (Sugawara et al. 1990; Ning et al. 1991; Ogata et al. 1993). Berube et al. (1998) have identified a senescence gene on chromosome 4 (termed MORF4) which selectively suppressed members of the B complementation group. Further information is not available at this time.

Sequences on the long arm of chromosome 6 (6q) have been specifically associated with immortalization of SV/HF. First, immortal SV/HF show karyotypic changes involving 6q when matched sets of pre-immortal and immortal SV/HF are compared (Hoffschir et al. 1992 Hubbard-Smith et al. 1992; Ray and Kraemer 1992). Secondly, we have shown that introduction of a normal chromosome 6 or 6q by microcell-mediated chromosome transfer suppresses growth of immortal SV/HF and induces a senescence-like appearance (Sandhu et al. 1994). We have proposed that genetic changes in a gene designated as *SEN6* is an essential step for immortalization of these SV/HF. Analysis of loss of heterozygosity (LOH) of 12 independent SV/HF from three different laboratories has defined a minimum deleted region at 6q27 (Banga et al. 1997). More recently, we have localized it at or distal to AF6 (MLLT4) in 6q27 (unpub. data). A YAC contig has been obtained for this region and representative YACs have been retrofitted with a marker suitable for testing for growth suppressor activity (Kim 1997). AF6 itself is a candidate gene for *SEN6* (Prasad et al. 1993). It is located within the suspect region. One allele is disrupted in an immortal SV/HF and genetic evidence supports a mutation in the other allele. The gene product has been localized to tight junctions (in rat epithelial cells) and has several interesting binding domains including ras (Khosravi-Far et al. 1996; Kuriyama et al. 1996) as well as an actin binding domain in an alternatively spliced mRNA (Mandai et al. 1997). It also has a PDZ domain seen in many proteins involved in protein-protein interaction, including the human tumor suppressor APC. Most interestingly, AF-6 has been recently shown to interact with the receptor protein-tyrosine kinase EphB3, a member of the largest sub-family of such surface receptors (Hock et al. 1998). Its role in replicative senescence and/or immortalization is unknown but one would expect it to be involved in one or more signal transduction pathways. We are in the process of developing definitive evidence on its role as *SEN6* in immortalization by seeking to determine whether there are mutations in AF6 which can alter or inactivate its functions. These results do not, however, preclude the existence of more than one growth suppressor on chromosome 6, even within 6q26–27. Rather, it emphasizes that one of these loci, i.e. *SEN6*, is involved in a particular stage of growth regulation, namely immortalization, and that such a locus is in this region. One might reasonably expect that human tumors would also be defective in *SEN6* since many tumors are immortal. In fact, chromosome 6 is among the most frequently rearranged chromosomes among human cancers (as reviewed in Mitelman 1991; Teyssier and Ferre 1992). More specifically, LOH involving 6q26–27 has been reported for non-Hodgkins lymphoma (Gaidano et al. 1992; Offit et al. 1993; Menasce et al. 1994), ovarian epithelial tumors (Foulkes et al. 1993; Wan et al. 1994; Saha et al. 1995; Saito et al. 1996) mammary tumors (Devilee et al. 1991; Orphanos et al. 1995), and mesotheliomas (Bell et al. 1997) although no gene has yet been identified as responsible for any of these tumors. Since more than one region on chromosome 6 has been found to have LOH in many of these studies, and growth suppression or reduced tumoricity by microcell-mediated chromosome transfer has been shown to involve 6q for melanoma cell lines (Trent et al. 1990), 6q14–21 for ovarian tumor cells (Sandhu et al. 1996), 6q21 for BK-virus-transformed mouse cells (Morelli et al. 1997), and 6q25–26 for the breast tumor cell line MCF-7 (Negrini et al. 1994), multiple growth suppressors may be present on 6q. Furthermore, the gene for Mn•SOD, which was previously noted to be reduced in immortal SV/HF, has been mapped to 6q25 (Church et al. 1992). Its overexpression inhibits tumor cell growth as well as SV/HF in culture (Church et al. 1993; Yan et al. 1996). Although the mechanism of suppression is not known, alteration of superoxide radicals could have wide-ranging effects including those on enzymes and transcription factors through redox-modulation as reviewed by Oberley (1998). Finally, additional loci may also be affected in SV40-immortalized cells as more extensive LOH on chromosome 6 is evident on prolonged passage. Mutations involving multiple loci affecting cell proliferation might be expected to be selected for by passage of immortal SV/HF under conventional culture conditions as reflected in increased efficiency of colony formation and shorter generation times (Neufeld et al. 1987), accumulation of chromosome rearrangements (Neufeld et al. 1987; Hubbard-Smith et al. 1992) and progressive lengthening of telomeric sequences (Small et al. 1996). Chromosome rearrangements involving chromosomes other than 6 are also to be expected and are routinely observed in this laboratory and others (Neufeld et al. 1987). Indeed, large T antigen facilitates karyotypic instability (Ray et al. 1990), at least in part through its effects on p53 and other checkpoint mediators. Such rearrangements may contribute to the development of immortal cell lines through changes in gene expression in addition to those relevant to the senescent phenotype.

5
Summary and Perspectives

In conclusion, immortalization involves changes in cell proliferation including overcoming effects of senescence and crisis. At least two stages can be defined. The first stage involves functions which are dependent on large T antigen and parallel many of the changes associated with transformed cells, whereas in the second stage additional changes independent of large T antigen play a critical role. Hence, one needs to consider SV40-mediated immortalization as a multi-faceted phenomenon. The availability of SV/HF which are not immortal, yet can give rise to immortal sublines is a valuable analytical tool:

(1) It involves inactivation of at least three growth suppressors pRB, p53 and *SEN6*. This requirement readily explains the inability to obtain spontaneous immortal human fibroblasts in culture. Less clear is why it is relatively easy to accomplish the goal in mouse fibroblasts, since roles for the same genes have been invoked in mouse cells.

(2) Telomere stabilization is needed. A TRF of at least a minimal length is likely to be required. The highly provocative finding with introduction of hTERT into human fibroblasts may explain multiple aspects but it becomes unclear why inactivation of pRB and p53 is needed to bypass replicative senescence. Is it simply due to the development of secondary blocks which now have to be overcome? Similarly, will introduction of hTERT into non-immortal SV/HF cause them to be immortal as predicted by the findings of Bodnar et al. (1998) or merely increase the likelihood for immortal cells to arise? It should also be noted that fluctuation in telomerase activity and TRF occurs in immortal tumor lines in culture, indicating it remains in a dynamic state in any case (Bryan et al. 1998).

(3) Non-immortal SV/HF need to undergo further changes which facilitate their ability to bypass apoptosis and crisis. The precipitating event of crisis is still unclear but may be related to the re-emergence of a senescence-like phenotype. It is not known whether senescent cells later

undergo apoptosis. Senescent human fibroblasts show enhanced resistance to apoptosis induced by deprivation of serum (Wang 1995) but this finding need not apply to other inducers of apoptosis. It is also plausible that changes which occur during extended life span of SV/HF (e.g. in the presence of inactivated pRB and p53 inter alia) set up a new paradigm for the cell, which might mimic those in multistage carcinogenesis *in vivo*.

(4) Ongoing efforts to identify changes in gene expression of SV/HF at the different steps of immortalization (i.e. extended life span, crisis, and recovery of immortal cells) utilizing global gene strategies hold the promise of providing new insights into primary and secondary effectors of immortalization as well as growth regulation.

Finally, these issues involving immortalization of SV40-transformed cells may have direct relevance to human carcinogenesis as recent studies have raised the issue of the presence of SV40 or SV40 sequences in several types of human tumors *in vivo* (Wiman and Klein 1997; Brown and Lewis 1998). Although the significance of such sequences, in general, and to carcinogenesis, in particular, is subject to debate, a key issue is whether these cells show an SV40-specific carcinogenic phenotype. At present, *SEN6* provides a rare, if not unique, criterion by which an SV40-transformed cell can be distinguished from one infected but not transformed (i.e. immortalized) by SV40. Analysis for specific changes described or proposed in this chapter may provide additional distinguishing criteria. It remains evident that cross fertilization of diverse experimental systems will continue to make an understanding of the SV40-cell interaction a source of insights into complex biological phenomena.

Acknowledgements. The author wishes to acknoweledge the valuable discussions and research contributions of past and present members of the laboratory, and the expert typing of Ms. Diane Muhammadi. Research cited from this laboratory was supported by the National Institutes of Health (US), National Institute of Aging grants AG04821 and AG00378.

References

Alcorta D, Xiong Y, Phelps D, Hannon G, Beach D, Barrett JC (1996) Involvement of the cyclin-dependent kinase inhibition p16 (INK4a) in replicative senescence of normal human fibroblasts. Proc Natl Acad Sci USA 93:13742–13747

Alfano C, McMacken R (1989) Heat shock protein-mediated disassembly of nucleoprotein structures is required for the initiation of bacteriophage lambda DNA replication. J Biol Chem 264:10709–10718

Allen RG (1998) Oxidative stress and superoxide dismutase in development, aging and gene regulation. AGE 21:47–76

Allsopp RC, Vaziri H, Patterson C, Goldstein S, Younglai EV, Futcher AB, Grieder CW, Harley CB (1992) Telomere length predicts replicative capacity of human fibroblasts. Proc Natl Acad Sci USA 89:10114–10118

Atadja P, Wong H, Garkavtsev I, Veillette C, Riabowol K (1995) Increased activity of p53 in senescing fibroblasts. Proc Natl Acad Sci USA 92:8348–8352

Banga SS, Kim S-H, Hubbard K, Dasgupta T, Jha KK, Patsalis P, Hauptschein R, Gamberi B, Dalla-Favera R, Kraemer P, Ozer HL (1997) SEN6, a locus for SV40-mediated immortalization of human cells, maps to 6q26–27. Oncogene 14:313–321

Bargonetti J, Reynisdottir I, Friedman P, Prives C (1992) Site specific binding of wild-type p53 to cellular DNA is inhibited by SV40 T antigen and mutant p53. Genes Dev 6:1886–1898

Bell DW, Jhanwar SC, Testa JR (1997) Multiple regions of allelic loss from chromosome arm 6q in malignant mesotheliomas. Cancer Res 57:4057–4062

Bertrand P, Rouillard D, Boulet A, Levalois C, Soussi T, Lopez B (1997) Increase of spontaneous intrachromosomal homologous recombination in mammalian cells expressing a mutant p53 protein. Oncogene 14:1117–1122

Berube NG, Smith JR, Pereira-Smith OM (1998) The genetics of cellular senescence. Am J Hum Genet 62:1015–1019

Bischoff FZ, Yim SO, Pathak S, Grant GMJ, Giovanclla B, Strong LC, Tainsky MA (1990) Spontaneous abnormalities in normal fibroblasts from patients with Li-Fraumeni cancer syndrome: aneuploidy and immortalization. Cancer Res 50:7979–7984

Blasco MA, Lee HW, Hande MR, Samper E, Lansdorp PM, DePinho RA, Greider CW (1997) Telomere shortening and tumor formation by mouse cells lacking telomerase RNA. Cell 91:25–34

Bodnar AG, Ouellette M, Frolkis M, Holt SE, Chiu C, Morin GB, Harley CB, Shay JW, Lichtsteiner S, Wright WE (1998) Extension of lifespan by introduction of telomerase into normal human cells. Science 279:349–352

Bravard A, Hoffschir F, Sabatier L, Ricoul M, Pinton A, Cassingena R, Estrade S, Luccioni C, Dutrillaux B (1992) Early superoxide dismutase alteration during SV40-transformation of human fibroblasts. Int J Cancer 52:797–801

Brehm A, Miska E, McCance D, Reid J, Bannister A, Kouzarides T (1998) Reintoblastoma protein recruits histone deacetylase to repress transcription. Nature 391:597–601

Brockman WW (1978) Transformation of Balb/c-3T3 cells by tsA mutants of simian virus 40: temperature sensitivity of transformed phenotype and retransformation by wild-type virus. J Virol 25:860–870

Brodsky JL, Pipas JM (1998) Polymavirus T antigens: molecular chaperones for multi- protein complexes. J Virol 72:5329–5334

Brown F, Lewis AM (1998) Simian virus 40 (SV40): a possible human polyomavirus. Dev Biol Stand 94:1–392

Brown JP, Wei W, Sedivy JM (1997) Bypass of senescence after disruption of p21 gene in normal diploid human fibroblasts. Science 277:831–834

Bryan TM, Englezou A, Gupta J, Bacchetti S, Reddel RR (1995) Telomere elongation in immortal human cells without detectable telomerase activity. EMBO J 14:4240–4248

Bryan TM, Englezou A, Dunham MA, Reddel RR (1998) Telomere length dynamics in telomerase-protein immortal human cell populations. Exp Cell Res 239:370–378

Caelles C, Helmberg A, Karin M (1994) p53-dependent apoptosis in the absence of transcriptional activation of p53-target genes. Nature 370:220–223

Campbell KS, Mullane KP, Ibraham IA, Stubdal H, Zalvide J, Pipas JM, Silver, PA, Roberts TM, Schauffhausen BS, DeCaprio JA (1997) DnaJ/hsp40 chaperone domain of SV40 large T antigen promotes efficient viral DNA replication. Genes Dev 11:1098–1110

Campisi J (1996) Replicative senescence: an old wives tale? Cell 84:497–500

Chang T, Ray FA, Thompson DNA, Schlegel R (1997) Disregulation of mitotic checkpoints and regulatory proteins following acute expression of SV40 large T antigen in diploid human cells. Oncogene 14:2383–2393

Chen S, Tsao Y, Chen Y, Huang S, Chang J, Wu S (1998) The induction of apoptosis by SV40 T antigen correlates with c-jun overexpression. Virology 244:521–529

Cheng R, Shammas M, Li J, Shmookler Reis RJ (1997) Expression of large T antigen stimulates reversion of a chromosomal gene duplication in human cells. Exp Cell Res 234:300–312

Chowdary DR, Dermody JJ, Jha KK, Ozer HL (1994) Accumulation of p53 in a mutant cell line defective in the ubiquitin pathway. Mol Cell Biol 14:1997–2008

Church ST, Grant J, Meese E, Trent JM (1992) Sublocalization of the gene encoding manganese superoxide dismutase (MnSOD/SOD2) to 6q25 by fluorescence *in situ* hybridization and somatic cell hybrid mapping. Genomics 14:823–825

Church ST, Grant JW, Ridnour LA, Oberley LW, Swanson P, Meltzer PS, Trent JM (1993) Increased manganese superoxide dismutase expression suppresses the malignant phenotype of human melanoma cells. Proc Natl Acad Sci USA 90:3113–3117

Cicala C, Avanaggiati M, Graessman A, Rundell K, Levine AS, Carbone M (1994) Simian virus 40 small-t antigen stimulates viral DNA replication in permissive monkey cells. J Virol 68:3138–3144

Cole C (1996) Polyomavirinae: the viruses and their replication. In: Fields, Knipe, Howley (eds) Fundamental Virology, 3rd edn. Lippincott-Raven, Philadelphia, Pennsylvania, pp 947–978

Conzen SD, Snay CA, Cole CN (1997) Identification of a novel antiapoptotic functional domain in simian virus 40 large T antigen. J Virol 71:4536–4543

Counter CM, Avilion AA, LeFeuvre CE, Stewart NG, Greider CW, Harley CB, Bacchetti S (1992) Telomere shortening associated with chromosome instability is arrested in immortal cells which express telomerase activity. EMBO J 11:1921–1929

Damania B, Alwine JC (1996) TAF-like function of SV40 large T antigen. Genes Dev 10:1369–1381

DeCaprio JA, Ludlow JW, Figge J, Shew JY, Huang CM, Lee WH, Marsilio E, Paucha E, Livinston DM (1988) SV40 large T antigen forms a specific complex with the product of the retinoblastoma susceptibility gene. Cell 54:275–283

DePinho RA (1998) The cancer-chromatin connection. Nature 391:533–536

Deppert W (1980) SV40 T-antigen-related surface antigens: correlated expression with nuclear T-antigen in cells transformed by an SV40 A-gene mutant. Virology 104:497–501

Deppert W, Haug M, Steinmayer T (1987) Modulation of p53 protein expression during cellular transformation with SV40. Mol Cell Biol 7:4453–4463

DeRonde A, Sol C, van Stein A, Schegget J, van der Noordaa J (1989) The SV40 small t antigen is essential for morphological transformation of human fibroblasts. Virology 171:260–263

Devilee P, van Vilet M, van Sloun P, Dijkshoorn NK, Hermans J, Pearson PL, Cornelisse CJ (1991) Allelotype of human breast carcinoma: a second major site for loss of heterozygosity is on chromosome 6q. Oncogene 6:1705–1709

Dimri GP, Lee X, Basile G, Acosta M, Scott G, Roskelley C, Medrano EE, Linskens M, Rubelj J, Pereira-Smith O, Peacocke M, Campisi J (1995) A biomarker that identifies senescent human cells in culture and in aging skin *in vivo*. Proc Natl Acad Sci USA 92:9363–9367

Dulic V, Drullinger L, Lees E, Reed S, Stein G (1993) Altered regulation of G1 cyclins in senescent human diploid fibroblasts: accumulation of inactive cyclin E-cdk2 and cyclin D1-cdk2 complexes. Proc Natl Acad Sci USA 90:11034–11038

Duncan EL, Whitaker NJ, Moy EL, Reddel RR (1993) Assignment of SV40- immortalized cells to more than one complementation group for immortalization. Exp Cell Res 205:337–344

El-Deiry WS, Tokino T, Velculescu V, Levy D, Parsons R, Trent J, Lin D, Mercer WE, Kinzlev KW, Vogelstein B (1993) WAF1, a potential mediator of p53 tumor suppression. Cell 75:817–825

Ferber A, Chang C, Sell C, Ptasznik A, Cristofalo VJ, Hubbard K, Ozer HL, Adamo M, Roberts CJ, LeRoith D, Dumenil G, Baserga R (1993) Failure of senescent human fibroblasts to express the insulin-like growth factor-1 gene. J Biol Chem 268:17883–17888

Foulkes WD, Ragoussis J, Stamp GWH, Allan GJ, Trowsdale J (1993) Frequent loss of heterozygosity on chromosome 6 in human ovarian carcinoma. Br J Cancer 67:551–559

Gaidano G, Hauptschein RS, Parsa NZ, Offit K, Rao PH, Lenoir G, Knowles DM, Chaganti RSK, Dalla-Favera R (1992) Deletions involving two distinct regions of 6q in B-cell non-Hodgkin lymphoma. Blood 80:1781–1787

Gershey E (1979) Simian virus 40-host cell interaction during lytic infection. J Virol 30:76–83

Giaccia A, Kastan M (1998) The complexity of p53 modulation: emerging patterns from divergent signals. Genes Dev 12:2973–2983

Girardi AJ, Jensen FC, Koprowski H (1965) SV40-induced transformation of human diploid cells: crisis and recovery. J Cell Comp Physiol 65:69–84

Giriat I, Schmidtt A, de Lange T (1998) Tankyrase, a poly (ADP-ribose) polymerase of human telomeres. Science 282:1484–1488

Gluzman Y, Sambrook J, Frisque RJ (1980) Expression of early genes of origin-defective mutants of SV40. Proc Natl Acad Sci USA 77:3898–3902

Goldstein S (1990) Replicative senescence: the human fibroblast comes of age. Science 249:1129–1133

Gorman SD, Cristofalo VJ (1985) Reinitiation of cellular DNA synthesis in BrdU-selected nondividing senescent WI-38 cells by SV40 infection. J Cell Physiol 125:122–126

Gruss C, Wetzel E, Baack M, Mock U, Knippers R (1988) High-affinity SV40 T-antigen binding sites on the human genome. Virology 167:349–360

Hara E, Tsurui H, Shinozaki A, Nakada S, Oda K (1991) Cooperative effect of antisense-Rb and antisense-p53 oligomers on the extension of lifespan in human diploid fibroblasts, TIG-1. Biochem Biophys Res Commun 179:528–534

Hara E, Smith R, Parry D, Tahara H, Stone S, Peters G (1996) Regulation of p16^{CDKN2} expression and its implications for cell immortalization and senescence. Mol Cell Biol 16:859–867

Harley CB, Futcher AB, Greider CW (1990) Telomere shortening during aging of human fibroblasts. Nature 346:866–868

Harper J, Adami G, Wei N, Keyomarssi K, Elledge SJ (1993) The p21 Cdk-interacting protein Cip1 is a potent inhibitor of G1 cyclin-dependent kinases. Cell 75:805–816

Harvey DM, Levine AJ (1991) p53 alteration is a common event in the spontaneous immortalization of primary BALB/c murine embryo fibroblasts. Genes Dev 5:2375–2385

Harvey M, Sands A, Weiss R, Hegi M, Wiseman R, Pantazis P, Giovanella B, Tainsky M, Bradley A, Donehower LA (1993) In vitro growth characteristics of embryo fibroblasts isolated from p53-deficient mice. Oncogene 8:2457–2467

Hayflick L (1965) The limited *in vitro* lifetime of human diploid cell strains. Exp Cell Res 37:614–636

Hess R, Brandner G (1997) DNA-damage-inducible p53 activity in SV40-transformed cells. Oncogene 15:2501–2504

Hinzpeter M, Deppert W (1987) Analysis of biological and biochemical parameters for chromatin and nuclear matrix association of SV40 large T antigen in transformed cells. Oncogene 1:119–129

Hock B, Bohme B, Karn T, Yamamoto T, Kaibuchi K, Holtrich U, Holland S, Pawson T, Rubsamen-Waigmann H, Strebhardt K (1998) PDZ-domain-mediated interaction of the Eph-related receptor tyrosine kinase EphB3 and the ras-binding protein AF6 depends on the kinase activity of the receptor. Proc Natl Acad Sci USA 95:9779–9784

Hoffschir F, Ricoul M, Lemieux N, Estrude S, Cassingena R, Dutrillaux B (1992) Jumping translocations originate clonal rearrangements in SV40-transformed human fibroblasts. Int J Cancer 52:130–136

Hoffschir F, Vuillaume M, Sabatier L, Ricoul M, Daya-Grosjean L, Estrade S, Cassingena R, Sarasin A, Dutrillaux B (1993) Decrease in catalase activity and loss of the 11p chromosome arm in the course of SV40 transformation of human fibroblasts. Carcinogenesis 14:1569–1572

Hubbard-Smith K, Patsalis P, Pardinas JR, Jha KK, Henderson AS, Ozer HL (1992) Altered chromosome 6 in immortal human fibroblasts. Mol Cell Biol 12:2273–2281

Ikram Z, Norton T, Jat PS (1994) The biological clock that measures the mitotic life-span of mouse embryo fibroblasts continues to function in the presence of simian virus 40 large tumor antigen. Proc Natl Acad Sci USA 91:6448–6452

Imai S, Takano T (1992) Loss of collagenase gene expression in immortalized clones of SV40 T antigen-transformed human diploid fibroblasts. Biochem Biophys Res Commun 189:148–153

Imai S, Fujino T, Nishibayashi S, Manabe T, Takano T (1994) Immortalization-susceptible elements and their binding factors mediate rejuvenation of regulation of the type I collagenase gene in simian virus 40 large T antigen-transformed immortal human fibroblasts. Mol Cell Biol 14:7182–7194

Kelley WL, Georgopoulos C (1997) The T/t common exon of simian virus 40, JC, and BK polyomavirus T antigens can functionally replace the J-domain of the *Escherichia coli* DnaJ molecular chaperone. Proc Natl Acad Sci USA 94:3674–3684

Khosravi-Far R, White MA, Westwick JK, Solski PA, Chrzanowska-Wodnicka M, Van Aest L, Wigler MH, Der CJ (1996) Oncogenic Ras activation of raf/mitogen-activated protein kinase-independent pathways is sufficient to cause tumorigenic transformation. Mol Cell Biol 16:3923–3933

Kierstead TD, Tevethia MJ (1993) Association of p53 binding and immortalization of primary C57B46 mouse embryo fibroblasts by using simian virus 40 T-antigen mutants bearing internal overlapping deletion mutations. J Virol 67:1817–1829

Kim S-H (1997) Localization of SEN6, a gene involved in SV40-mediated immortalization of human cells. Thesis dissertation. UMDNJ (GSBS), Newark, NJ

Kim NW, Piatyszek MA, Prowse KR, Harley CB, West MD, Ho PLC, Coviello GM, Wright WE, Weinrach SL, Shay JW (1994) Specific association of human telomerase activity with immortal cells and cancer. Science 266:2011–2015

Ko L, Prives C (1996) p53: puzzle and paradigm. Genes Dev 10:1054–1072

Kuriyama M, Harada N, Kuroda S, Yamamoto T, Nakafuka M, Iwamatsu A, Yamamoto D, Prasad R, Croce C, Canaani E, Kaibuchi K (1996) Identification of AF-6 and canoe as putative targets for Ras. J Biol Chem 27:607–610

LaBella F, Ozer HL (1985) Differential replication of SV40 and polyoma DNA in Chinese hamster ovary cells. Virus Res 2:329–343

Lenahan M (1998) Analysis of crisis in SV40-transformed human firbroblasts. Thesis dissertation. UMDNJ (GSBS), Newark, NJ

Lenahan MK, Ozer HL (1996) Induction of c-myc mediated apoptosis in SV40-transformed rat fibroblasts. Oncogene 12:1847–1854

Lill NL, Grossman S, Ginsberg D, DeCaprio J, Livingston D (1997a) Binding and modulation of p53 by p300/CBP coactivators. Nature 387:823–827

Lill NL, Tevethia, MJ, Eckner R, Livingston D, Modjtahedi N (1997b) p300 family members associate with the carboxyl terminus of SV40 large tumor antigen. J Virol 71:129–137

Lin JY, Simmons DJ (1991) The ability of large T-antigen to complex with p53 is necessary for the increased lifespan and partial transformation of human cells by simian virus 40. J Virol 65:6447–6453

Loeken M, Bikel I, Livingston D, Brady J (1988) Trans-activation of RNA polymerase II and III promoters by SV40 small t antigen. Cell 55:1171–1177

Lundblat V, Blackburn EH (1993) An alternate pathway for yeast telomere maintenance rescues est1⁻ senescence. Cell 73:347–360

Ma L, Broomfield S, Lavery C, Lin S, Xiao W, Bacchetti S (1998) Up-regulation of CIR1/CROC1 expression upon cell immortalization and in tumor-derived human cell lines. Oncogene 17:1321–1326

Maclean K, Rogan E, Whitaker N, Chang A. Rowe P, Dalla-Pozza L, Symonds G, Reddel RR (1994) *In vitro* transformation of Li-Fraumeni syndrome fibroblasts by SV40 large T antigen mutants. Oncogene 9:719–725

Magnaghi-Jaulin L, Groisman R, Nagwbneva I, Robin P, Lorain S, LeVillain J, Troalen F, Trouche D, Harel-Bellan A (1998) Retinoblastoma protein represses transcription by recruiting a histone deacetylase. Nature 391:601–604

Mandai K, Nakanishi H, Satoh A, Obaishi H, Wada M, Nishioka H, Itoh M, Mizoguchi A, Aoki T, Fujimoto T, Matsuda Y, Tsukita S, Takai Y (1997) Afadin: A novel actin filament-binding protein with one PDZ domain localized at cadherin-based cell-to-cell adherens junction. J Cell Biol 139:517–528

McCormick JJ, Maher VM (1988) Towards an understanding of the malignant transformation of diploid human fibroblast. Mutat Res 199:273–291

Mekeel K, Tang W, Kachnic L, Luo C, DeFrank J, Powell S (1997) p53 suppresses homologous recombination. Oncogene 14:1847–1857

Menasce LP, Orphanos V, Santibanex-Koref M, Boyle JM, Harrison CJ (1994) Common region of deletion on the long arm of chromosome 6 in non-Hodgkin's lymphoma and acute lymphoblastic leukaemia. Genes Chromosome Cancer 10:286–288

Mitelman F (1991) Catalog of chromosome aberrations in cancer. 4th edn, vol 1, Wiley-Liss, New York, pp 393–474

Miyashita T, Reed JC (1995) Tumor suppressor p53 is a direct transcriptional activator of the human bax gene. Cell 80:293–299

Moorwood K, Price T, Mayne V (1996) Mutation of p53 is not a prerequisite for immortalization of human fibroblats by SV40 T antigen. Exp Cell Res 223:308–313

Morelli C, Sherratt T, Trabanelli C, Rimessi P, Gualandi F, Greaves MJ, Negrini M, Boyle JM, Barbanti-Brodano G (1997) Characterization of a 4-Mb region at chromosome 6q21 harboring a replicative senescence gene. Cancer Res 57:4153–4157

Mungre S, Enderle K, Tuok B, Porras A, Wu Y, Muby M, Rundell R (1994) Mutations which affect the inhibition of protein phosphastase 2A by SV40 small t-antigen *in vitro* decrease viral transformation. J Virol 68:1675–1681

Negrini M, Sabbioni S, Possati L, Rattan S, Corallini A, Barbanti-Brodano G, Croce C (1994) Suppression of tumorigenicity of breast cancer cells by microcell-mediated chromosome transfer: studies on chromosomes 6 and 11. Cancer Res 54:1331–1336

Neufeld DS, Ripley S, Henderson A, Ozer HL (1987) Immortalization of human fibroblasts transformed by origin-defective SV40. Mol Cell Biol 7:2794–2802

Ning Y, Weber JL, Killary AM, Ledbetter DH, Smith JR, Pereira-Smith OM (1991) Genetic analysis of indefinite division in human cells: evidence for a cell senescence-related gene(s) on human chromosome 4. Proc Natl Acad Sci USA 88:5635–5639

Noda A, Ning Y, Venable SF, Pereira-Smith OM, Smith JR (1994) Cloning of senescent cell-derived inhibitors of DNA synthesis using an expression screen. Exp Cell Res 211:90–98

Oberley LW (1998) Inhibition of tumor cell growth by overexpression of manganese–containing superoxide dismutase. AGE 21:95–98

O'Connor R, Kauffman-Zeh A, Liu Y, Lehar S, Evan GI, Baserga R, Blattner WA, (1997) Identification of domains of the insulin-like growth factor I receptor that are required for protection from apoptosis. Mol Cell Biol 17:427–435

Offit K, Parsa NZ, Gaidano G, Filippa DA, Louie D, Pan D, Jhanwar SC, Dalla-Favera R, Chaganti RSK (1993) 6q deletions define distinct clinico-pathologic subsets of non-Hodgkin's lymphoma. Blood 82:2157–2162

Ogata T, Ayusawa D, Namba M, Takahasi E, Oshimura M, Oishi M (1993) Chromosome 7 suppresses indefinite division of nontumorigenic immortalized human fibroblast cell lines KMST-6 and SVSM-1/. Mol Cell Biol 13:6036–0643

O'Neill FJ, Hu Y, Chen T, Carney H (1997) Identification of p53 unbound to T-antigen in human cells transformed by SV40 T-antigen. Oncogene 14:955–966

V, McGown G, Hey Y, Boyle JM, Santibanez-Kore M. (1995) Proximal 6q, a region showing allele loss in primary breast cancer. Br J Cancer 71:290–293

Ozer HL, Slater ML, Dermody JJ, Mandel M (1981) Replication of SV40 DNA in normal human fibroblasts and in fibroblasts from xeroderma pigmentosum. J Virol 39:481–489

Ozer HL, Banga SS, Dasgupta T, Houghton J, Hubbard K, Jha KK, Kim S-H, Lenahan M, Pang Z, Pardinas JR, Patsalis P (1996) SV40-mediated immortalization of human fibroblasts. Exp Gerontol 31:303–310

Pages J, Manteuil S, Stehelin D, Fiszman M, Marx M, Girard M (1973) Relationship between replication of SV40 DNA and specific events of the host cell cycle. J Virol 12:99–107

Pardinas J, Pang Z, Houghton J, Palejwala V, Donnelly R, Hubbard K, Small MB, Ozer HL (1997) Differential gene expression in SV40-mediated immortalization of human fibroblasts. J Cell Physiol 171:325–335

Pereira-Smith OM, Smith JR (1981) Expression of SV40 T antigen in finite lifespan hybrids of normal and SV40-transformed fibroblasts. Somatic Cell Genet 7:411–421

Pereira-Smith OM, Smith JR (1983) Evidence for the recessive nature of cellular immortality. Science 221:964–966

Pereira-Smith OM, Smith JR (1988) Genetic analysis of indefinite division in human cells: identification of four complementation groups. Proc Natl Acad Sci USA 85:6042–6046

Peterson S, Gradbois D, Bradbury E, Kraemer PM (1995) Immortalization of human fibroblasts by SV40 large T antigen results in the reduction of cyclin D1 expression and subunit association with proliferating cell nuclear antigen and Waf1. Cancer Res 55:4651–4657

Porras A, Bennett J, Howe A, Tokos K, Bouck N, Henglein B, Sathyamangalam S, Thimmapaya B, Rundell K (1996) A novel simian virus 40 early-region domain mediates transactivation of the cyclin A promoter by small-t antigen and is required for transformation in small-t antigen-dependent assays. J Virol 70:6902–6908

Prasad R, Gu Y, Adler H, Nakamura T, Canaani O, Sato H, Huebner K, Gale RP, Nowell PC, Kuriyama K, Miyazaki Y, Croce CM, Canaani E (1993) Cloning of the ALL-1 fusion partner, the AF-6 gene, involved in acute myeloid leukemia with the t (6;11) chromosome translocation. Cancer Res 53:5624–5628

Qin XQ, Livingston DM, Ewen E, Sellers WR, Adams PD (1994) Deregulated E2F-1 transcription factor expression leads to S-phase entry and p53-mediated apoptosis. Proc Natl Acad Sci USA 91:10918–10922

Radna RL, Caton Y, Jha KK, Kaplan P, Li G, Traganos F, Ozer HL (1989) Growth of immortal simian virus 40 tsA-transformed human fibroblasts is temperature dependent. Mol Cell Biol 9:3093–3096

Ray FA, Kraemer PM (1992) Frequent deletions at chromosome 6q21 and other recurrent changes in nine newly immortalized human fibroblast cell lines. Cancer Genet Cytogenet 59:39–44

Ray FA, Peabody DS, Cooper JL, Cram LS, Kraemer PM (1990) SV40 T antigen alone drives karyotypic instability that preceeds neoplastic transformation of human diploid fibroblast. J Cell Biochem 13–31

Resnick-Silverman L, Pang Z, Li G, Jha KK, Ozer HL (1991) Retinoblastoma protein and simian virus 40-dependent immortalization of human fibroblasts. J Virol 65:2845–2852

Robinson CG, Pipas JM (1998) SV40 large tumor antigen (T antigen): database of mutants. Nucleic Acids Res 26:295–296

Rubelj I, Pereira-Smith OM (1994) SV40-transformed human cells in crisis exhibit changes that occur in normal cellular senescence. Exp Cell Res 211:82–89

Saffer JD, Jackson SP, Thurston SJ (1990) SV40 stimulates expression of the transacting factor Sp1 at the mRNA level. Genes Dev 4:659–666

Saha V, Lillington DM, Shelling AN, Chaplin T, Yaspo M, Ganesan TS, Young BD, (1995) AF6 gene on chromosome band 6q27 maps distal to the minimal region of deletion in epithelial ovarian cacer. Genes Chromosomes Cancer 14: 220–222

Saito S, Sirahama S, Matsushima M, Suzuki M, Sagae S, Kudo R, Saito J, Noda K, Nakamura Y (1996) Definition of a commonly deleted region in ovarian cancers to a 300-kb segment of chromosome 6q27. Cancer Res 56:5586–5589

Sandhu AK, Hubbard K, Kaur GP, Jha KK, Ozer HL, Athwal RS (1994) Senescence of immortal human fibroblasts by the introduction of normal human chromosome 6. Proc Natl Acad Sci USA 91:5498–5502

Sandhu AK, Kaur GP, Reddy DE, Rane NS, Athwal RS (1996) A gene on 6q14–21 restores senescence to immortal ovarian tumor cells. Oncogene 12:247–252

Satoh Y, Kishimura M, Kaneko S, Karasaki Y, Higashi K, Gotoh S (1994) Cloning of cDNAs with possible association with senescence and immortalization of human cells. Mutat Res 316:25–36

Sawai ET, Butel JS (1989) Association of a cellular heat shock protein with simian virus 40 large T antigen in transformed cells. J Virol 63:3961–3973

Scheffner M, Werness B, Huibregtse J, Levine A, Howley PM (1990) The E6 oncoprotein encoded by human papillomavirus types 16 and 18 promotes the degradation of p53. Cell 63:1129–1136

Scheidtmann KH, Mumby MC, Rundell K, Walter G (1991a) Dephosphorylation of simian virus 40 large-T antigen and p53 protein by protein phosphatase 2A: inhibition by small-t antigen. Mol Cell Biol 11:1996–2003

Scheidtmann KH, Virshup DM, Kelly TJ (1991b) Protein phosphatase 2A dephosphorylates SV40 large T antigen specifically at residues involved in regulation of the DNA- binding activity. J Virol 65:2098–2101

Schenker T, Trueb B (1998) Down-regulated proteins of mesenchymal tumor cells. Exp Cell Res 239:161–168

Schenker T, Lach C, Kessler B, Calderara S, Trueb B (1994) A novel GTP-binding protein which is selectively repressed in SV40 transformed fibroblasts. J Biol Chem 269:25447–25453

Sedivy JM (1998) Can ends justify the means? Telomeres and the mechanism of replicative senescence and immortalization in mammalian cells. Proc Natl Acad Sci USA 95:9078–9081

Serrano M, Lin AW, McCurrach ME, Beach D, Lowe SW (1997) Oncogenic ras provokes premature cell senescence associated with accumulation of p53 and p16. Cell 88:593–602

Shay JW, Wright WE (1989) Quantitation of the frequency of immortalization of normal diploid fibroblasts by SV40 large T-antigen. Exp Cell Res 1874:109–118

Shikama N, Lyon L, La Thanque NB (1997) The p300/CBP family: integrating signals with transcription factors and chromatin. Trends Cell Biol 7:230–236

Slansky JE, Li Y, Kaelin WG, Farnham PJ (1993) A protein synthesis-dependent increase in E2F1 mRNA correlates with growth regulation of the dihydrofolate reductase promoter. Mol Cell Biol 13:1610–1618

Small MB, Gluzman Y, Ozer HL (1982) Enhanced transformation of human fibroblasts by origin-defective simian virus 40. Nature 296:671–672

Small MB, Hubbard K, Pardinas J, Marcus AM, Dhanaraj SM, Sethi KA (1996) Maintenance of telomeres in SV40-transformed pre-immortal and immortal human fibroblasts. J Cell Phys 168:727–736

Srinivasan A, McClellan AJ, Vartikar J, Marks I, Cantalupo P, Li Y, Whyte P, Rundell K, Brodsky JL, Pipas JM (1997) The amino-terminal transforming region of simian virus 40 large T and small t antigens functions as a J domain. Mol Cell Biol 17:4761–1773

Stadlbauer F, Voitenleiter C, Bruckner A, Fanning E, Nasheuer H (1996) Species-specific replication of SV40 DNA *in vitro* requires the p180- subunit of human DNA polymerase α-primase. Mol Cell Biol 16:94–104

Stamps AC, Gusterson BA, O'Hare MJ (1992) Are tumors immortal? Eur J Cancer 28A:1495–1500

Stein G (1985) SV40-transformed human fibroblasts: Evidence for cellular aging in precrisis cells. J Cell Physiol 125:36–44

Stein GH, Besson M, Gordon L (1990) Failure to phosphorylate retinoblastoma gene product in senescent human fibroblasts. Science 249:666–669

Stein GH, Drullinger LF, Robetorye RS, Pereira-Smith OM, Smith JR (1991) Senescent cells fail to express cdc2, cycA and cycB in response to mitogen stimulation. Proc Natl Acad Sci USA 88:11012–11016

Stubdal H, Zalvide J, DeCaprio JA (1996) Simian virus 40 large T antigen alters the phosphorylation state of the RB-related proteins p130 and p107. J Virol 70:2781–2788

Stubdal H, Zalvide J, Campbell K, Schweitzer C, Roberts T, DeCaprio JA (1997) Inactivation of pRB-related proteins p130 and p107 mediated by the J domain of SV40 large T antigen. Mol Cell Biol 17:4979–4990

Sugawara O, Oshimura M, Koi M, Annab LA, Barrett JC (1990) Induction of cellular senescence in immortalized human cells by human chromosome I. Science 247:707–710

Symonds H, Krall L, Remington L, Saenz-Robles M, Lowe S, Jacks T, Van Dyke T (1994) p53-dependent apoptosis suppresses tumor growth and progression *in vivo*. Cell 78:703–711

Takekawa M, Maeda T, Saito H (1998) Protein phosphatase 2Cα inhibits the human stress-responsive p38 and JNK MAPK pathways. EMBO J 17:4744–4752

Tara Y (1997) Rb kinases and RB-binding proteins: new points of view. Trends Biochem Sci 22:14–17

Tevethia MJ, Lacko HA, Conn A (1998) Two regions of simian virus 40 large T-antigen independently extend the life span of primary C57BL/6 mouse embryo fibroblasts and cooperate in immortalization. Virology 243:303–312

Teyssier JR, Ferre D (1992) Identification of a clustering of chromosomal breakpoints in the analysis of 203 human primary solid tumor non specific karyotypic rearrangements. Anticancer Res 12: 997–1004.

Tiemann F, Zerrahn J, Deppert W (1995) Cooperation of SV40 large and small T antigens in metabolic stabilization of tumor suppressor p53 during cellular transformation. J Virol 69:6115–6121

Tooze J (ed) (1981) DNA tumor viruses, 2nd edn. Cold Spring Harbor Laboratory. Cold Spring Harbor , New York

Trent JM, Stanbridge EJ, McBride HL, Meese EU, Casey G, Araujo DE, Witkowski CM, Nagle RB (1990) Tumorigenicity in human melanoma cell lines controlled by introduction of human chromosome 6. Science 247:568–571

Tresini M, Cristofalo VJ (1998) Defects in MAPK-mediated signal transduction during replicative senescence of human fibroblasts. In: Bohr V, Clark B, Stevnsner T, Svejgaard A (eds) Molecular biology of aging. Munksgaard International, Copenhagen, Denmark

Van Steensel B, de Lange T (1997) Control of telomere length by the human telomeric protein TRF1. Nature 385:470–473

Vellucci VF, Germino FJ, Reiss M (1995) Cloning of putative growth regulatory genes from primary human keratinocytes by subtractive hybridization. Gene 106:213–220

Wan M, Zweizig S, D=Ablaing G, Zheng J, Velicescu M, Dubean L (1994) Three distinct regions of chromosome 6 are targets of loss of heterozygosity in human ovarian carcinomas. Int J Oncol 5:1043–1048

Wang E (1995) Senescent human fibroblasts resist programmed cell death and failure to suppress bcl-2 is involved. Cancer Res 55:2284–2292

Watanabe G, Howe A, Lee R, Albanese C, Shu I, Karneziss A, Zon L, Kyriakis J, Rundell K, Pestell RG (1996) Induction of cyclin D1 by SV40 small tumor antigen. Proc Natl Acad Sci USA 93:12861–12866

Weinberg R (1995) The Rb protein and cell cycle control. Cell 81:323–330

West MD, Pereira-Smith OM, Smith JR (1989) Replicative senescence of human skin fibroblasts correlates with a loss of regulation and overexpression of collagenase activity. Exp Cell Res 184:138–147

White E (1996) Life, death, and the pursuit of apoptosis. Genes Dev 10:1015

Wiessmuller L, Cammenga J, Deppert WW (1996) In vivo assay of p53 function in homologous recombination between SV40 chromosomes. J Virol 70:737–744

Wiman KG, Klein G (1997) An old acquaintance resurfaces in human mesothelioma. Nat Med 3:839–840

Wright WE, Pereira-Smith OM, Shay JW (1989) Reversible cellular senescence: a two-stage model for the immortalization of normal diploid fibroblasts. Mol Cell Biol 9:3088–3092

Xiong Y, Zhang H, Beach D (1993) Subunit rearrangement of the cyclin-dependent kinases is associated with cellular transformation. Genes Dev 7:1572–1583

Yaciuk P, Carter M, Pipas J, Moran E (1991) SV40 large T antigen expresses a biological activity complementary to the p300-associated transforming function of the adenovirus E1A gene products. Mol Cell Biol 11:2116–2124

Yan T, Oberley LW, Zhong W, St. Clair DK (1996) Manganese-containing superoxide dismutase overexpression causes phenotypic reversion in SV40-transformed human lung fibroblasts. Cancer Res 56:2864–2871

Yang S-I, Lickteig RL, Estes R, Rundell K, Walter G, Mumby MC (1991) Control of protein phosphatase 2A by simian virus 40 small-t antigen. Mol Cell Biol 11:1988–1995

Yew P, Liu X, Berk AJ (1994) Adenovirus E1B oncoprotein tethers a transcriptional repression domain to p53 genes. Genes Dev 8:190–202

Zalvide J, Stubdal H, DeCaprio JA (1998) The J domain of SV40 large T antigen is required to functionally inactivate RB family proteins. Mol Cell Biol 18:1408–1415

Zerrahn J, Knippschild U, Winkler T, Deppert W (1993) Independent expression of the transforming amino-terminal domain of SV40 large T antigen from an alternatively spliced third SV40 early mRNA. EMBO J 12:4739–4746

Zhai W, Tuan J, Comai L (1997) SV40 large T antigen binds to the TBP-TAF I complex SL1 and coactivates ribosomal RNA transcription. Genes Dev 11:1605–1617

Zhu J, Rice PW, Chamberlan M, Cole CN (1991) Mapping the transcriptional transactivation function of SV40 large T-antigen. J Virol 65:2778–2790

Zouzias D, Jha KK, Mulder C, Basilico C, Ozer HL (1980) Transformation of human fibroblasts by SV40 DNA. Virology 104:439–453 the Keller & Gehring (1992) for 800 bp sequence is the EMBL database accession number Z12001 (Keller & Gehring, personal communication, Data Submitted to...

Chu-Lagraff Q., and Doe C.Q. (1993). Neuroblast specification and formation regulated by wingless in the Drosophila CNS. Science 1, 100–108.

Bopp D. T., V. Madhavan, Kimble J., Cech T.R. (1991). Posttranscriptional regulation of sex-determination in Drosophila.

Concepts of Immortalization in Human Mammary Epithelial Cells

K. Swisshelm[1]

1
Introduction

Cellular immortality is a hallmark of neoplastic transformation, the detection of which in vitro may be only limited by providing stringent culture conditions for neoplastic cells. Unlike cultured rodent cells, spontaneously immortal human mammary epithelial cell lines cultured from presumably normal cells have rarely been identified. Only two reports described cell lines that arose in mammary tissue explants obtained from women with fibrocystic disease or intraductal hyperplasia (Briand et al. 1987; Soule et al. 1990). For clarity, and by definition, a normal human mammary epithelial cell (HMEC) population or clone that eventually senesces in culture is termed a "strain", while a clone or population that has unlimited replicative potential is termed a "line". The principles and players that confer unlimited in vitro replicative potential upon HMEC strains overlap partially with findings from cultured human diploid lineages, e.g., fibroblast-like cells and keratinocytes. Similarities between immortalized fibroblast-like cultures and HMECs include the functional repression of cell cycle- and DNA damage responsive-proteins p53, pRB, p16, and p21 (reviewed in Campisi et al. 1996 ; Garkavtsev et al. 1998). Nonetheless, human HMECs have unique phenotypes and require stringent culture conditions. This chapter will be limited to a discussion of human-derived mammary epithelial cells.

Breast cancer is the most frequently diagnosed female cancer in first world countries and the major cause of death in women between the ages of 40 and 55 years (Couch and Weber 1998). The molecular events responsible for benign breast diseases, a heterogeneous group of lesions, is remarkably limited, given the high incidence. Two benign lesions, hyperplasia and fibrocystic disease, arise in epithelial components of breast tissue in mid to late life and may contain precancerous cells (Schnitt et al. 1991). Benign breast disease affects over 7 million females over age 50 in the US alone (Vogel 1991) and carries a relative risk of ≥ 2 for developing into breast cancer, particularly in premenopausal women (Schnitt et al. 1991 ; London et al. 1992).

[1] Department of Pathology Box 357470, 1959 N.E. Pacific Street University of Washington Seattle, Washington 98195–7470 USA.

Progress in Molecular and Subcellular Biology, Vol. 24
A. Macieira-Coelho
© Springer-Verlag Berlin Heidelberg 1999

The enormity of this public health issue underscores the importance of study of the early events in mammary cell biology, which permit progression to detectable cancer and unrestricted proliferation.

2
Cultured HMECs

2.1
Establishment of Primary Cell Strains

A number of defined and semi-defined culture systems have been developed to support the growth of primary HMECs (reviewed in Smith 1991). Initial attempts to culture mammary cells began with human milk fluids, and the first successful culture of large numbers of primary HMECs was accomplished by Hammond, Stampfer and colleagues wherein enzymatic digestion of reduction mammoplasty tissue served as the primary inoculum of organoid HMEC. Cellular outgrowth from the primary explants was favored by the development of specialized media, with or without a low percentage of fetal calf sera and/or the inclusion of bovine pituitary extracts (Stampfer et al. 1980 ; Hammond et al. 1984 ; Stampfer and Bartley 1985). Later developments in medium formula for primary and long-term HMEC culture have included modifications in the basal salts (Band and Sager 1989) and a minimal concentration of calcium (Soule and McGrath 1986). For an extensive treatise on the history and issues of primary HMEC culturing, see the HMEC homepage: http://www.lbl.gov/~mrgs.

Primary HMEC cultures are obtained from reduction mammoplasties, mastectomies or lactational fluids, and the "cobblestone"-appearing cells of heterogeneous morphology exhibit unique markers, for example, expression of a specific subset of intermediate filaments, keratins and polymorphic epithelial mucin. The HMEC-specific proteins are found in cells of both the luminal and basal lineage in the mammary gland in vivo. The range of intermediate filament and mucin proteins expressed in cultured cells suggests that they arise from a common stem cell precursor or from both luminal and basal precursors (Chang and Taylor-Papadimitriou 1983 ; Taylor-Papadimitriou et al. 1989 ; Bartek et al. 1990).

2.2
Life Span of HMEC Strains

Similar to all normal human somatic cells, HMEC strains have a limited in vitro replicative life span potential and, under specific culture conditions, can be trypsinized and passaged more than twenty times at various split ratios, yielding up to 100 population doublings. Little consistent data is available on the possible correlation of donor age and number of HMEC population doublings. Recent data from 124 individuals from the Baltimore Longitudinal Study of Aging finds that in vitro replicative life span of explanted dermal fibroblasts is not correlated with donor age. This latter study suggests that parameters that influence culture life span include the health status of the donor. Moreover, tissues from which primary cultures are derived likely consist of a patchwork of "young" and "old" lineages (Cristofalo et al. 1998). I suggest that correlating donor age with replicative life span of HMECs would require careful attention to variables such age of menarche, number of menstrual cycles, and levels of circulating hormones during in vivo replication.

2.3
HMEC Senescence

As HMECs approach the end of their in vitro replicative life span, they exhibit variable and complex changes, which include a flattened morphology with enlarged cytoplasmic volume and increased vacuolization. Consistent with findings detected with in vitro and in vitro senescent fibroblasts and keratinocytes (Dimri et al. 1995), a cultured population of senescent HMECs are positive for the pH 6.0-dependent staining for β-galactosidase activity (Brenner et al. 1998), a biomarker associated with replicative cellular senescence. Brenner and colleagues determined that elevated expression of the cyclin dependent kinase inhibitor protein, p16, accompanies the acquisition of the senescent phenotype, a landmark finding which is consistent with loss of p16 function in immortal HMECs (Brenner et al. 1998, see Sect. 6).

One study, comparing gene expression between senescent HMECs and senescent keratinocytes, uncovered similar phenotypes from the two different cell types. Saunders and colleagues detected diminished expression in both kinds of cells for two genes associated with proliferation, CDC2 and E2F-1, as well as an increase in expression for a squamous cell differentiation gene, cornifin (small proline-rich protein 1a) at approximately 38 population doublings (Saunders et al. 1993). E2F is a key protein in the G_1-S cell cycle transition phase, transcriptionally regulating several genes which determine competence to proceed through the cell cycle, including pRB and cyclin E (Weinberg 1996). CDC2 is a subunit of the cell cycle promoting factor responsible for G_1-S and G_2-M transitions, universal among eukaryotes (O'Connor 1997). These latter investigations with keratinocytes and HMECs illustrate a corollary: pathways to immortalization may also jettison a differentiation program, and the state of in vitro cellular senescence may represent a subset of terminally differentiated cells. However, a degree of caution is required in summarily equating "senescence" and terminal differentiation in cultured HMECs. Under certain culture conditions, such as withdrawal of growth factors, elevated calcium, or alteration of extracellular matrix conditions (addition of collagen I and/or absence of fibronectin), HMECs undergo an apparent squamous-type differentiation (Soule and McGrath 1986; Satoh et al. 1993).

3
Means and Mechanisms of HMEC Immortaliztion

Immortal HMECs established from primary, limited life span cultures have been derived using a variety of strategies, including treatment with chemical carcinogens (Stampfer and Bartley 1985); SV40 large tumor (T) antigen (Bartek et al. 1991 ; Shay et al. 1993); introduction of human papilloma virus DNA (Band et al. 1990); γ-irradiation (Wazer et al. 1994); as well as spontaneous derivatives from individuals with Li-Fraumeni syndrome (Shay et al. 1995). The mechanism for the last two studies are likely to be due to a loss of p53 function, via a primary heterozygous state of a p53 mutation or loss of an allele following DNA damage, respectively. Immortal HMECs are not "malignant." They are unable to form colonies in semi-solid growth conditions, which leads to anchorage independent growth, and they fail to produce tumor growth in nude mice.

3.1
Chemically Induced Immortalization

The first in vitro immortalized human epithelial cells were established from cultured HMECs following treatment with the chemical carcinogen benzo[a]pyrene (B[a]P), wherein only two indefinite life span lines arose (Stampfer and Bartley 1985). It is noteworthy that HMECs are able to metabolize B[a]P, which may contribute to the generation of oxidative species such as thymine glycols (Leadon et al. 1988) and bulky adducts, which suggests that chemical carcinogenesis could occur in the mammary gland. Moreover, extracts present in mammary lipids may contribute to single strand-breaks in HMEC DNA (Martin et al. 1997). The B[a]P-immortalized HMECs continue to display epithelial–specific markers similar to the finite life span normal cultures, such as keratin 5 and 14, but not keratin 19 (Taylor-Papadimitriou et al. 1989; Stampfer and Yaswen 1992). The B[a]P-immortalized HMECs cannot form colonies on methocel or produce tumors in nude mice (Stampfer and Bartley 1985). However, further transformation can be achieved if the cells are transduced with a retrovirus containing murine oncogenic genes, such as *v-ras*, and to a lesser extent, the *mos* gene (Clark et al. 1988). The immortal cultures can also be transformed and are tumorigenic when provided with either mutated or elevated levels of the cell surface receptor, ERB-B2 (Pierce et al. 1991). ERB-B2 (Her-2/NEU) is amplified or over-expressed in over 30 % of primary human breast cancers (Slamon et al. 1989; Clark and McGuire 1991). The ERB-B2 gene encodes a growth factor receptor with tyrosine kinase activity, a component of intracellular signaling. It has been suggested that cell signaling in immortal or tumorigenic HMECs is deranged with abnormal levels of ERB-B2, as deduced by the concomitant induction of the cellular phosphotases, LAR and PTP1B (Zhai et al. 1993).

3.2
DNA Tumor Viruses

3.2.1
SV40 virus

SV40 tumor virus is not etiologic for human cancers, but SV40 functions have been invaluable in identifying and understanding components of transcriptional regulation and of the mammalian cell cycle. A breast cancer cell line (HBL-100) that was thought to be a spontaneously immortal isolate from human milk, was later found to contain integrated, tandemly duplicated SV40 genome, with defects in the SV40 DNA replication functions (Caron de Fromentel et al. 1985; Saint-Ruf et al. 1989). Similar to studies with the B[a]P-immortalized HMECs, HBL-100 cells are non-tumorigenic. HBL-100 can be transformed by transfection with the oncogenic v-Kirsten-ras gene (Saint-Ruf et al. 1988).

Experimental models demonstrate that SV40 viral DNA or presence of the SV40 large T antigen results in rare HMEC immortalization (Bartek et al. 1991 ; Lebeau et al. 1995). Human milk cells immortalized with SV40 large T antigen are not able to form colonies on a semi-solid medium. However, a tumorigenic line was derived following further introduction of a mutant Harvey-ras gene (Bartek et al. 1991). Chang and colleagues attempted immortalization of the "type I and II" HMECs by SV40 viral DNA transfection and determined that the type I (luminal) cells were more susceptible (Kao et al. 1997). Shay, Wright, and co-investigators determined that the frequency of HMEC immortalization is elevated compared to fibroblast cells using SV40 T antigen (Shay et al. 1993a). The relative ease of HMEC immortalization by SV40 T antigen may be the result of differential post selection for HMECs which may have loss or altered key suppressor function, e.g., loss of p16. Both the large and small T-antigen proteins can bind and potentially eliminate the function of the tumor suppressor, p53. The SV40 oncoproteins may mechanistically impinge on other cellular functions during HMEC immortalization. SV40 T-antigen can bind to cell cycle complexes, including p107, cyclinA/cdk2, and cyclin E/cdk2 (reviewed in Ludlow and Skuse 1995). Moreover, both large and small T-antigens interact with basal transcription factors as evidenced in many cells types, with the potential to activate heterologous promoters, e.g., cyclin A, B, and D (reveiwed in Moens et al. 1997). Findings in other cell systems suggest that multiple mechanisms of SV40 gene action could alter the host cell cycle regulatory functions during HMEC immortalization.

3.2.2
Human Papilloma Virus

Human papilloma virus (HPV) DNA, types 16 and 18, are potent HMEC immortalizing agents. Band, Sager and colleagues first demonstrated that transfection with DNA from HPV E6 or HPV E6+E7, but not E7 alone produced immortal clones and cultures (Band et al. 1990). At the time of these original studies, neither p53 nor p16 status had been determined. Like SV40 and carcinogen-immortalized HMECs, HPV DNA-immortalized HMECs are non-tumorigenic. The postulated mechanism for the immortal phenotype directed by HPV E6 is via inactivation of functions of the cellular p53 protein. The HPV E6 protein forms a complex with p53 which leads to a ubiquitin-mediated degradation of p53 protein (Band et al. 1991 ; Xu et al. 1995 ; Dalal et al. 1996). Initially it was hypothesized that inactivation of both p53 function (by HPV E6 protein) and pRB (by HPV E7 protein) were required for robust HMEC immortalization. However, more recent studies employing retroviral transduction of HPV genes reveal additional insights. Late passage HMECs are preferentially immortalized by HPV E6, while early HMEC explants can be immortalized by HVP E7 alone, as assessed by the unique keratin profiles for early and later passages of HMEC cultures (Wazer et al. 1995). Moreover, cells derived from human milk, either early or late passage, required both HPV 16 E6 and E7 for extended life (Wazer et al. 1995). These additional findings of Band and co-workers suggest that specific HMEC lineages are susceptible to a unique spectrum of oncogenic insults and/or loss of tumor suppressor functions leading to immortalization. An alternative interpretation comes from studies by Foster and Galloway. Expanding upon criteria developed by Shay and colleagues, who describe the unique sensitivity of HMECs, in comparison to fibroblasts, to immortalization by HPV 16 E6, and the requirement for by passing two levels of growth restriction, termed "M1 and M2" (Shay et al. 1993b), Foster and Galloway suggest that HMECs must first "bypass" a quiescent state which is facilitated by HPV E7. Thereupon, HMECs are able to quickly overcome the normally lengthy time required to perpetuate a population of extended life cells and never acquire the large cell morphology (Foster and Galloway 1996).

3.3
Other Avenues to Immortalization?

In order to obtain immortal HMECs, cellular damage or interference with homeostatic functions in replication and transcription is required, which suggests that cultured human glandular mammary epithelial cells, like cultured human fibroblast-like cells have a highly regulated and recently evolved means to limit life span as observed in culture. HMT-3522 and MCF-10 are two independent lines of "spontaneously" immortalized HMECs, and these lines were derived from mammoplasty tissue from individuals with fibrocystic disease. The HMT-3522 line arose from benign fibrocystic tissue of a 50-year-old woman (Briand et al. 1987). The MCF-10 line was obtained from a 36-year-old woman with fibrocystic disease (Soule et al. 1990) and is wild-type for p53 (Katayose et al. 1995). Although fibrocystic disease by itself does not carry a elevated risk for the development of atypical hyperplasia or breast cancer, it is possible that the culture conditions provide the appropriate environment for immortal clones to emerge from fibrocystic epithelial components.

Perhaps the most novel and serendipitous HMEC immortalization recorded came from accidental heat shock (45 °C) due to a defective tissue culture incubator (MCF-12 cells). Interestingly, this MCF-12 cell line was originally derived from mammoplasty tissue from an individual with contra-lateral infiltrating ductal carcinoma (Paine et al. 1992 ; Wolman et al. 1994).

Endogenous, heterozygous p53 mutation is an alternative route for obtaining immortal, cultured HMECs. Four of nine HMEC cultures from a woman in a Li-Fraumeni pedigree (M133T, with a high penetrance for breast cancer) underwent in vitro immortalization. Attempts to establish immortal lines from mammary tissue fibroblasts from the same individual failed (Shay et al. 1995) perhaps because the fibroblasts failed to override the p16/RB pathway which appears to be essential for epithelial cell immortalization (Kiyono et al. 1998; Weinberg 1998).

4
Cytogenetic Aberrations in Immortal HMECs

Cytogenetic analysis of immortalized HMECs has revealed that most cells that survive the selective insults, such as carcinogen-induced immortalization or inactivation of wild-type p53, are near diploid or hypodiploid, with distinct clonal chromosome abnormalities. Among the lines examined from different immortalization strategies, only a few consistent hallmarks have been detected, indicating that if there is a common theme in gain or loss of genetic material required for HMEC-specific immortalization, it cannot be resolved at the light microscope level. Cytogenetic analyses are limited to acquisition of sufficient metaphase cells. Therefore, the identification of early landmarks of genomic rearrangement during the initial establishment of the immortal HMEC cultures is seldom feasible. However, with the combined technologies of spectral karyotyping (Veldman et al. 1997) and competitive genomic hybridization (Forozan et al. 1997), this shortfall need not be an issue in the future.

One of the first cytogenetic analyses was from the two B[a]P-induced immortal HMEC clones, which were followed for more than 40 passages. The extended life cultures, prior to immortalization, were of a normal, diploid female complement, 46,XX. Upon immortalization, the two lines exhibited one to four abnormalities each, including, in one line, the deletion of part of the long arm of one chromosome 3, loss of a whole chromosome 6, and deletion of the long arm of one chromosome 12. Moreover, the acquired aberrations observed in the B[a]P-immortal lines were stable, which is in contrast to most cultured solid breast tumor lines. By passage 41–47, the two lines had acquired an additional two to six clonal aberrations (Walen and Stampfer 1989).

Cytogenetic analysis of HPV-immortalized HMECs revealed that most clones were near diploid or hypodiploid, with both structural and numerical aberrations. However, all lines, whether the cells were immortalized by the entire HPV 16 or 18 genome, or the HPV16 E6 DNA alone, or in combination with E7 DNA, exhibited a preferential loss of one chromosome 19, within the earliest possible time of analysis at passage 4 (Swisshelm et al. 1992). The specific loss of chromosome 19 suggests that a gene or genes on chromosome 19 confers regulates replicative potential. It is noteworthy that TGFβ1 (Sects. 5.2 and 6.1) maps to chromosome 19.

Cytogenetically, clones from the spontaneously immortalized MCF-10 line exhibited a normal female diploid karyotype for approximately 2 years, whereupon a translocation between the short arm of 3 and the short arm of chromosome 9 was detected, t(3;9)(p13;p22) (Soule et al. 1990; Wolman et al. 1994). At the time this first abnormality was observed, the investigators noted more rapid growth and the emergence of viable floating cells. Both the floating and attached cells gave rise to sublines with multiple structural and numerical aberrations. The most striking and consistent abnormality was loss of one chromosome 20 (Wolman et al. 1994). Loss of chromosome 20 was observed as the second most frequent whole chromosome loss in HPV-immortalized HMECs (Swisshelm et al. 1992). The other spontaneous immortal line, HMT-3522, was first karyotyped at passage 16, after 10 months in culture, and exhibited the following alterations: the loss of a whole chromosome 6, and derivative chromosomes 8, 13, 14 and 17 (Nielsen and Briand 1989 ; Nielsen et al. 1994). It is notable that loss of a whole chromosome 6 was also observed as one of the earliest aberrations in the 184 A1 B[a]P-immortalized HMECs (Walen and Stampfer 1989). The HMT-3522 line has been cultured for more 13 years and has continued to evolve karyotypically, with additional whole chromosome loss, translocations, and, by passage 205 the development of a hyper-triploid side-line. The original investigators revisited passage 45 HMT-3522 cells using molecular cytogenetic methods to verify their earlier findings and both confirmed and revised the earlier interpretations to include amplification of the c-MYC locus on chromosome 8 and partial loss of material on the short arm of chromosome 1 (band 1p35) (Nielsen et al. 1997). Loss of material from chromosome one and amplification of the c-MYC gene are both alterations detected in primary breast cancer.

Karyotyping of HMECs immortalized by SV40 revealed overlapping themes of chromosome aberrations compared with those detected in other immortalization schemes. Again, different lineages exhibited different stem-line abnormalities (Lebeau et al. 1995). Moreover, one SV40 hypodiploid line showed loss of an X chromosome, which was the first chromosome abnormality to be detected in the B[a]P-immoral HMECs (Walen and Stampfer 1989). Loss of an X chromosome was also detected in an MCF-10 cell sideline (Wolman et al. 1994) and in one HPV 18-immortalized HMEC line (Swisshelm et al. 1992). Remarkably, one of the SV40 lines became tumorigenic and was able to support tumor formation in nude mice (Lebeau et al. 1995). The tumorigenic phenotype was correlated with the karyotypically determined loss of the short arm of one chromosome 3. Loss of heterozygosity of the 3p, particularly 3p24 is frequently observed in breast cancer as well as in surrounding normal tissue (Deng et al. 1996).

In summary, immortal HMECs may exhibit either gross or minimal chromosome instability, dependent upon the mechanism of immortalization, but in most cases the lines are near diploid. The detection of near diploid karyotypes in immortal HMECs contrasts with karyotypic evaluations of primary or metastatic breast cancer lines, which are highly aneuploid.

5
Growth Factors and Cytokines: Positive and Negative Regulators

5.1
EGF and TGFα

The growth factor requirements of normal and immortal HMECs has been previously reviewed (Stampfer and Yaswen 1992). Key factors necessary for establishment, maintenance and cloning of HMECs include: insulin, epidermal growth factor (EGF), and hydrocortisone as well as undefined hormones and factors present in fetal calf sera or bovine pituitary extracts. The key growth factors necessary for selection and maintenance of HMEC cultures are different from those necessary for fibroblast cultures, which require high percentages of fetal calf sera (10–16%) that contains platelet derived growth factor.

EGF or transforming growth factor alpha (TFGα), which activates the same high affinity cell surface receptor, EGF-receptor (EGF-R), are salient HMEC growth factors, as demonstrated in experiments in which the action of EGF was attenuated by withdrawal of EGF and by addition of receptor inhibitory antibodies (Stampfer et al. 1993a). Removal of EGF alone from the culture medium of both normal and conditionally immortal HMEC resulted in a reversible block in G_0/G_1 of the cell cycle; after addition of EGF both normal and immortal cells were able to synchronously re-enter the cell cycle. After fully blocking cycling cells by both a receptor antibody and withdrawal of EGF followed by release with re-addition of EGF, normal HMECs showed a rapid induction of early response genes, c-MYC, c-FOS and c-JUN. In contrast, immortal HMECs never lose expression of these early response genes during the cell cycle block, but do exhibit a modest up-regulation when EGF is reconstituted (Stampfer et al. 1993a).

Carcinogen treated immortal HMECs show less stringent growth factor requirements. Rare cell lines from *N*-nitros-ethyl-urea (ENU)-treated cells, B[a]P- immortal HMECs, can be propagated in the absence of EGF, bovine pituitary extract, or insulin. Nonetheless, the ENU-treated cells lack the ability to support clonal growth in semi-soft medium or to form tumors in nude mice (see review in Stampfer and Yaswen 1992).

Between 50,000 and 100,000 high affinity EGF receptors are found on normal HMECs (Bates et al. 1990; Berthon et al. 1992). Briand and colleagues have demonstrated that long-term passage (greater than 118 passages) of the spontaneously immortal HMEC line, HMT-3522, which is p53 negative, results in a sub-line with the ability to grow in the absence of EGF (Moyret et al. 1994). Concomitant with EGF-independence, these investigators observed a modestly enhanced mRNA expression of TGFα and the EGF receptor. These two factors are likely to be responsible for an autocrine stimulation of the cells. Following continuous culture for an additional 120 passages, the EGF-independent cells gave rise to a line which was tumorigenic, with the only karyotypic abnormality being the addition of an extra short arm of chromosome 7, a region in which the EGFR maps (Madsen et al. 1992). These studies suggest that selection for growth factor independence followed, by inactivation of p53 or genetic instability may propagate tumorigenicity.

5.2
TGFβ

Normal HMEC growth is inhibited by the multi-functional peptide, transforming growth factor beta-1 (TGFβ). Conditionally immortal B[a]P-treated HMECs cultured with TGFβ gradually become growth inhibited. However, immortal cells resistant to the effects of TGFβ can be obtained, and they are fully immortal (Stampfer and Yaswen 1992). The mechanism of proliferative stasis imposed by TGFβ is probably not accounted for by the TGFβ-induced expression of extracellular matrix (ECM) proteins or ECM-remodeling proteins. TGFβ up-regulates the expression of fibronectin, collagen IV, laminin B1, collagenase, urokinase type plasminogen activator, and plasminogen activator inhibitor 1 in both normal and immortal HMECs (Stampfer et al. 1993b). Alterations in gene expression of ECM proteins appear to be independent of TGFβ's growth inhibitory effects, as both the TGFβ-sensitive and -partially resistant HMECs up-regulated the ECM genes (Stampfer et al. 1993). It is currently hypothesized that the TGFβ-resistance phenotype acquired during a gradual conversion to immortality is the result of an epigenetic phenomenon, including critical shortening of telomeres (Stampfer et al. 1997).

6
Cell Cycle Alterations

6.1
TGFβ and the INK4a-ARF Complex

What are the mechanisms of the TGFβ-induced growth inhibition? Two lines of evidence indicate that TGFβ affects the levels of cell cycle proteins, through an unidentified signal transduction cascade. Activation or up-regulation of the inhibitors of cyclin D-dependent kinases appear to be the most immediate targets. p15^{INK4B} (p15) and p16^{INK4A} (p16) are part of the INK4a-ARF genomic locus (multiple tumor suppressor locus 1 and 2), which contains three differentially spliced proteins. p16 is strongly implicated as a tumor suppressor gene for many cell types, and the p16 gene product is up-regulated in cellular senescence (reviewed in Haber 1997).

Through examination of TGFβ-induced growth arrest in normal and immortal HMECs, it was determined that p15 expression is induced and the protein is stabilized by addition of cytokine. The precise protein interaction(s) with p15 in HMECs have not been clearly delineated, but probably involve the cdk-cyclinD complex. These stable complexes are lost in TGFβ-resistant immortal HMECs. Sandhu and colleagues also determined that only p15 was stabilized through a TGFβ-mediated pathway (Sandhu et al. 1997).

6.2
p16 and Telomerase

p16 is a likely critical partner in mammary replicative homeostasis, and inactivation of p16 may subsequently lead to unregulated cell proliferation of HMECs. Although loss of pRB may not be required for HMEC immortalization, loss of p16 function is likely to be required, along with additional genotypic or phenotypic alterations. The inactivation of the p16/RB pathway may occur in response to external "stresses" (reviewed in Weinberg 1998). In two independent strategies for HMEC immortalization – chemical carcinogenesis and HPV 16 E6 DNA transduction – p16 function was abrogated through a silencing mechanism involving DNA cytosine methylation (Brenner et al. 1998 ; Foster et al. 1998). Huschtscha et al. have observed loss of p16 expression accompanied by cytosine DNA methylation of p16 within the core CpG island in extended life (not immortal) HMEC cultures. The loss of expression was specific for HMECs, since mammary-derived fibroblasts from the same individuals did not show the same DNA alterations (Huschtscha et al. 1998). An earlier report found that HPV16 E6 readily activated telomerase in early passage HMECs or keratinocytes, but not in fibroblasts, and that the activation was independent of inactivation of p53 (Klingelhutz et al. 1996). Is the inactivation of p16 and the activation of telomerase the keys to HMEC immortalization?

Telomerase activity is low or undetectable in human somatic cells and telomere shortening is characteristic of cultured senescent fibroblasts. Introduction of the human telomerase reverse transcriptase subunit (hTRT) extends the life span of fibroblasts as well as retinal pigmented epithelial cells, and eliminates the pH6-dependent B-galactosidase senescence biomarker (Bodnar et al. 1998). As discussed by Sedivy, activation of telomerase may be the salient event required for immortalization (Sedivy 1998), although it remains to be fully tested in human HMECs. Recently, Kiyono and colleagues provided evidence that reconstitution of the catalytic component of telomerase can immortalize only HMECs that have lost p16 expression (Kiyono et al. 1998).

Activation of telomerase has been detected in immortal HMECs that have acquired a gradual resistance to TGFβ, a process termed, "conversion" (Stampfer et al. 1997). Thus, with either the rapid induction of immortalization by HPV E6 or the gradual conversion of HMECs by selection for TGBβ resistance, a similar phenotype is obtained with respect to telomerase activity. Moreover, there may exist two different mechanisms for maintenance of telomere length: one activated by near-senescent cells to repair critically short telomeres. Perhaps a different mechanism is utilized during rapid HPV E6 immortalization. How do the in vitro findings correlate with breast carcinoma? The clinical data is not definitive, but suggests that telomerase activity is present in primary cancer, including ductal carcinoma in situ. In two recent studies, 106 patient samples from stage 1-IV breast cancer were analyzed for telomerase, and the activity was correlated with expression of cyclins D, E, pRB and p53. Approximately 85 % of the primary breast tumors were telomerase positive, but large variations in the levels of activity were apparent. Notably, there exists a strong association between higher cyclin D1 or E levels and telomerase activity and, in contrast, a more favorable prognosis with low telomerase activity (Landberg et al. 1997; Roos et al. 1998). Interestingly, primary breast tumors with the highest telomerase activity had the lowest detectable p16 protein (Landberg et al. 1997). A more recent report compared telomerase activity and expression with breast cancer histopathology, from benign to invasive, and found significant correlation between high telomerase activity and severity of disease (Yashima et al. 1998). In conclusion, activation of telomerase impinges upon both HMEC immortalization and the drive towards malignancy and invasive carcinoma.

7
Perspectives

Immortalization is the acquired phenotypic and genotypic change necessary to overcome replicative senescence. Compared with human fibroblasts, HMECs can be immortalized with a higher frequency via SV40- or inactivation of p53-mediated pathways (Shay et al. 1993). In vitro senescence likely reflects a normal in vivo homeostatic state responsible for organismal size, growth, repair and suppression of oncogenesis (Sager 1991). Multiple insults to normal cultured HMECs have led to the isolation of immortal cultures which undergo successive rounds of genetic alterations, as evidenced by clonal cytogenetic aberrations, including gross loss, gains, and translocations.

Is HMEC immortalization a linear process? Are the first steps a culmination of loss of repression, impinging upon pRB function, and do the second steps affect p53 function? Is telomere erosion, leading to activation of telomerase, required for an irreversible escape from senescence and "true" HMEC replicative immortality? Experimental data argue for this model.

In reviewing the model systems for immortalization, it would be remiss to state that cultured HMECs, normal or fully immortal, reflect the in vivo differentiated function of mammary cells, such as lactation or involution, or a complete model of the primary events in human breast cancer. Breast cancer cells, especially from early cancers, are difficult to adapt to conventional culture systems (Smith et al. 1987; Dairkee et al. 1995; Saxena et al. 1997). Moreover, breast cancer cytogenetic analysis reveals extreme intra-tumor heterogeneity (Teixeira et al. 1996). However, immortal HMECs provide a useful model system for elaborating early events in neoplastic transformation. Many of the abnormalities seen in immortal HMEC are also seen in breast cancer, for example, loss of all or a portion of the short arm of chromosome 3, amplification or duplication of chromosome 8, including amplification of the cMYC gene, and trisomy 20 or amplification of a region of 20. Immortal HMEC are likely to sustain an inherent chromosome instability (CIN), evidenced by a continued karyotypic evolution in culture, which is consistent with the evolving pattern detected in primary breast carcinomas. Lengauer and colleagues suggest that in most human cancers chromosome instability is a dominant trait (Lengauer et al. 1997; 1998). Most early passage immortal HMECs are near diploid and, in time, evolve complex karyotypes including chromosome loss, gain and structural rearrangements. There is no evidence for telomere fusions due to possible telomere shortening even in the earliest passage, but this could reflect lethality or loss of the cells which could undergo such events during the selection process.

Are conditionally immortal HMECs, then, closer to the cancer phenotype or the "normal?" This issue has been addressed in the experiments of Zajchowski and co-workers, wherein tumor and immortal cells were fused. The hybrids displayed most of the characteristics of the immortal cells, including keratin expression and the failure to form tumors in nude mice (Zajchowski et al. 1990). The molecular alterations leading to HMEC immortalization are currently converging upon several key molecules – p53, p16 and telomerase – implicating a convergence between cell cycle and chromatin integrity. The future challenge will be to magnify the road map of immortalization events to include the earliest molecular and patho-physiological landmarks.

Acknowledgment. A special thanks goes to Dr. Martha Stampfer for her stimulating and thoughtful discussions during the preparation of this manuscript.

References

Band V, Sager R (1989) Distinctive traits of normal and tumor-derived human mammary epithelial cells expressed in a medium that supports long-term growth of both cell types. Proc Natl Acad Sci USA 86:1249–1253

Band V, Zajchowski D, Kulesa V, Sager R (1990) Human papilloma virus DNAs immortalize normal human mammary epithelial cells and reduce their growth factor requirements. Proc Natl Acad Sci USA 87:463–467

Band V, De Caprio JA, Delmolino L, Kulesa V, Sager R (1991) Loss of p53 protein in human papillomavirus type 16 E6-immortalized human mammary epithelial cells. J Virol 65:6671–6676

Bartek J, Bartkova J, Taylor-Papadimitriou J (1990) Keratin 19 expression in the adult and developing human mammary gland. Histochem J 22:537–544

Bartek J, Bartkova J, Kyprianou N, Lalani EN, Staskova Z, Shearer M, Chang S, Taylor-Papadimitriou J (1991) Efficient immortalization of luminal epithelial cells from human mammary gland by introduction of simian virus 40 large tumor antigen with a recombinant retrovirus. Proc Natl Acad Sci USA 88:3520–3524

Bates SE, Valverius EM, Ennis BW, Bronzert DA, Sheridan JP, Stampfer MR, Mendelsohn J, Lippman ME, Dickson RB (1990) Expression of the transforming growth factor-alpha/epidermal growth factor receptor pathway in normal human breast epithelial cells. Endocrinology 126:596–607

Berthon P, Pancino G, de Cremoux P, Roseto A, Gespach C, Calvo F (1992) Characterization of normal breast epithelial cells in primary cultures: differentiation and growth factor receptors studies. In Vitro Cell Dev Biol 28A:716–724

Bodnar AG, Ouellette M, Frolkis M, Holt SE, Chiu CP, Morin GB, Harley CB, Shay JW, Lichtsteiner S, Wright WE (1998) Extension of life-span by introduction of telomerase into normal human cells. Science 279:349–352

Brenner AJ, Stampfer MR, Aldaz CM (1998) Increased p16 expression with first senescence arrest in human mammary epithelial cells and extended growth capacity with p16 inactivation. Oncogene 17:199–205

Briand P, Petersen OW, Van Deurs B (1987) A new diploid nontumorigenic human breast epithelial cell line isolated and propagated in chemically defined medium. In Vitro Cell Dev Biol 23:181–188

Campisi J, Dimri G, Hara E (1996) Control of replicative senescence. In: Schneider EL, Rowe JW (eds) Handbook of the biology of aging. Academic Press, San Diego, pp 121–149

Caron de Fromentel C, Nardeux PC, Soussi T, Lavialle C, Estrade S, Carloni G, Chandrasekaran K, Cassingena R (1985) Epithelial HBL-100 cell line derived from milk of an apparently healthy woman harbours SV40 genetic information. Exp Cell Res 160:83–94

Chang SE, Taylor-Papadimitriou J (1983) Modulation of phenotype in cultures of human milk epithelial cells and its relation to the expression of a membrane antigen. Cell Differ 12:143–154

Clark GM, McGuire WL (1991) Follow-up study of HER-2/neu amplification in primary breast cancer. Cancer Res 51:944–948

Clark R, Stampfer MR, Milley R, O'Rourke E, Walen KH, Kriegler M, Kopplin J, McCormick F (1988) Transformation of human mammary epithelial cells by oncogenic retroviruses. Cancer Res 48:4689–4694

Couch FJ Weber BL (1998) Breast cancer. In: Vogelstein B, Kinzler KW (eds) The genetic basis of human cancer. McGraw-Hill, New York, pp 537–563

Cristofalo VJ, Allen RG, Pignolo RJ, Martin BG, Beck JC (1998) Relationship between donor USA 95:10614–10619

Dairkee SH, Deng G, Stampfer MR, Waldman FM, Smith HS (1995) Selective cell culture of primary breast carcinoma. Cancer Res 55:2516–2519

Dalal S, Gao Q, Androphy EJ, Band V (1996) Mutational analysis of human papillomavirus type 16 E6 demonstrates that p53 degradation is necessary for immortalization of mammary epithelial cells. J Virol 70:683–688

Deng G, Lu Y, Zlotnikov G, Thor AD, Smith HS (1996) Loss of heterozygosity in normal tissue adjacent to breast·carcinomas. Science 274:2057–2059

Dimri GP, Lee X, Basile G, Acosta M, Scott G, Roskelley C, Medrano EE, Linskens M, Rubelj I, Pereira-Smith O, Peacocke M, Campisi J (1995) A biomarker that identifies senescent human cells in culture and in aging skin in vivo. Proc Natl Acad Sci USA 92:9363–9367

Forozan F, Karhu R, Kononen J, Kallioniemi A, Kallioniemi OP (1997) Genome screening by comparative genomic hybridization. Trends Genet 13:405–409

Foster SA, Galloway DA (1996) Human papillomavirus type 16 E7 alleviates a proliferation block in early passage human mammary epithelial cells. Oncogene 12:1773–1779

Foster SA, Wong DJ, Barrett MT, Galloway DA (1998) Inactivation of p16 in human mammary epithelial cells by CpG island methylation. Mol Cell Biol 18:1793–1801

Garkavtsev I, Hull C, Riabowol K (1998) Molecular aspects of the relationship between cancer and aging: tumor suppressor activity during cellular senescence. Exp Gerontol 33:81–94

Haber DA (1997) Splicing into senescence: the curious case of p16 and p19ARF. Cell 91:555–558

Hammond SL, Ham RG, Stampfer MR (1984) Serum-free growth of human mammary epithelial cells: rapid clonal growth in defined medium and extended serial passage with pituitary extract. Proc Natl Acad Sci USA 81:5435–5439

Huschtscha LI, Noble JR, Neumann AA, Moy EL, Barry P, Melki JR, Clark SJ, Reddel RR (1998) Loss of p16INK4 expression by methylation is associated with lifespan extension of human mammary epithelial cells. Cancer Res 58:3508–3512

Kao CY, Oakley CS, Welsch CW, Chang CC (1997) Growth requirements and neoplastic transformation of two types of normal human breast epithelial cells derived from reduction mammoplasty. In Vitro Cell Dev Biol Anim 33:282–288

Katayose D, Gudas J, Nguyen H, Srivastava S, Cowan KH, Seth P (1995) Cytotoxic effects of adenovirus-mediated wild-type p53 protein expression in normal and tumor mammary epithelial cells. Clin Cancer Res 1:889–897

Kiyono T, Foster SA, Koop JI, McDougall JK, Galloway DA, Klingelhutz AJ (1998) Both Rb/p16INK4a inactivation and telomerase activity are required to immortalize human epithelial cells. Nature 396:84–88

Klingelhutz AJ, Foster SA, McDougall JK (1996) Telomerase activation by the E6 gene product of human papillomavirus type 16. Nature 380:79–82

Landberg G, Nielsen NH, Nilsson P, Emdin SO, Cajander J, Roos G (1997) Telomerase activity is associated with cell cycle deregulation in human breast cancer. Cancer Res 57:549–554

Leadon SA, Stampfer MR, Bartley J (1988) Production of oxidative DNA damage during the metabolic activation of benzo[a]pyrene in human mammary epithelial cells correlates with cell killing. Proc Natl Acad Sci USA 85:4365–4368

Lebeau J, Gerbault-Seureau M, Lemieux N, Apiou F, Calvo F, Berthon P, Goubin G, Dutrillaux B (1995) Loss of chromosome 3p arm differentiating tumorigenic from non- tumorigenic cells derived from the same SV40-transformed human mammary epithelial cells. Int J Cancer 60:244–248

Lengauer C, Kinzler KW, Vogelstein B (1997) Genetic instability in colorectal cancers. Nature 386:623–627

Lengauer C, Kinzler KW, Vogelstein B (1998) Genetic instabilities in human cancers. Nature 396:643–649

London SJ, Connolly JL, Schnitt SJ, Colditz GA (1992) A prospective study of benign breast disease and the risk of breast cancer. J Am Med Assoc 267:941–944

Ludlow JW, Skuse GR (1995) Viral oncoprotein binding to pRB, p107, p130, and p300. Virus Res 35:113–121

Madsen MW, Lykkesfeldt AE, Laursen I, Nielsen KV, Briand P (1992) Altered gene expression of c-myc, epidermal growth factor receptor, transforming growth factor-alpha, and c-erb-B2 in an immortalized human breast epithelial cell line, HMT-3522, is associated with decreased growth factor requirements. Cancer Res 52:1210–1217

Martin FL, Venitt S, Carmichael PL, Crofton-Sleigh C, Stone EM, Cole KJ, Gusterson BA, Grover PL, Phillips DH (1997) DNA damage in breast epithelial cells: detection by the single-cell gel (comet) assay and induction by human mammary lipid extracts. Carcinogenesis 18:2299–2305

Moens U, Seternes OM, Johansen B, Rekvig OP (1997) Mechanisms of transcriptional regulation of cellular genes by SV40 large T- and small T-antigens. Virus Genes 15:135–154

Moyret C, Madsen MW, Cooke J, Briand P, Theillet C (1994) Gradual selection of a cellular clone presenting a mutation at codon 179 of the p53 gene during establishment of the immortalized human breast epithelial cell line HMT-3522. Exp Cell Res 215:380–385

Nielsen KV, Briand P (1989) Cytogenetic analysis of in vitro karyotype evolution in a cell line established from nonmalignant human mammary epithelium. Cancer Genet Cytogenet 39:103–118

Nielsen KV, Madsen MW, Briand P (1994) In vitro karyotype evolution and cytogenetic instability in the non-tumorigenic human breast epithelial cell line HMT-3522. Cancer Genet Cytogenet 78:189–199

Nielsen KV, Niebuhr E, Ejlertsen B, Holstebroe S, Madsen MW, Briand P, Mouridsen HT, Bolund L (1997) Molecular cytogenetic analysis of a nontumorigenic human breast epithelial cell line that eventually turns tumorigenic: validation of an analytical approach combining karyotyping, comparative genomic hybridization, chromosome painting, and single-locus fluorescence in situ hybridization. Genes Chromosomes Cancer 20:30–37

O'Connor PM (1997) Mammalian G1 and G2 phase checkpoints. Cancer Surv 29:151–182

Paine TM, Soule HD, Pauley RJ, Dawson PJ (1992) Characterization of epithelial phenotypes in mortal and immortal human breast cells. Int J Cancer 50:463–473

Pierce JH, Arnstein P, DiMarco E, Artrip J, Kraus MH, Lonardo F, Di Fiore PP, Aaronson SA (1991) Oncogenic potential of erbB-2 in human mammary epithelial cells. Oncogene 6:1189–1194

Roos G, Nilsson P, Cajander S, Nielsen NH, Arnerlov C, Landberg G (1998) Telomerase activity in relation to p53 status and clinico-pathological parameters in breast cancer. Int J Cancer 79:343–348

Sager R (1991) Senescence as a mode of tumor suppression. Environ Health Perspect 93:59–62

Saint-Ruf C, Nardeux P, Estrade S, Brouty-Boye D, Lavialle C, Rhim JS, Cassingena R (1988) Accelerated malignant conversion of human HBL-100 cells by the v-Ki-ras oncogene. Exp Cell Res 176:60–67

Saint-Ruf C, Nardeux P, Cebrian J, Lacasa M, Lavialle C, Cassingena R (1989) Molecular cloning and characterization of endogenous SV40 DNA from human HBL-100 cells. Int J Cancer 44:367–372

Sandhu C, Garbe J, Bhattacharya N, Daksis J, Pan CH, Yaswen P, Koh J, Slingerland JM, Stampfer MR (1997) Transforming growth factor beta stabilizes p15INK4B protein, increases p15INK4B-cdk4 complexes, and inhibits cyclin D1-cdk4 association in human mammary epithelial cells. Mol Cell Biol 17:2458–2467

Satoh H, Sawada N, Watanabe Y, Satoh M, Hirata K, Mori M (1993) Human mammary epithelial cells undergo squamous differentiation in serum-free three-dimensional culture upon loss of growth activity. Cell Struct Funct 18:315–321

Saunders NA, Smith RJ, Jetten AM (1993) Regulation of proliferation-specific and differentiation-specific genes during senescence of human epidermal keratinocyte and mammary epithelial cells. Biochem Biophys Res Commun 197:46–54

Saxena S, Jain AK, Pandey KK, Dewan AK (1997) Role of steroid hormone and growth factor receptors and proto-oncogenes in the behavior of human mammary epithelial cancer cells in vitro. Pathobiology 65:75–82

Schnitt S, Connolly JL, Scalafani L, Smith BL, Morrow M, Eberlein TJ (1991) Benign breast disorders. In: Harris JR, Hellman S, Henderson IC and Kinne DW (eds) Breast Diseases. JB Lippincott, Philadelphia, pp 15–50

Sedivy JM (1998) Can ends justify the means? Telomeres and the mechanisms of replicative senescence and immortalization in mammalian cells. Proc Natl Acad Sci USA 95:9078–9081

Shay JW, Van Der Haegen BA, Ying Y Wright WE (1993a) The frequency of immortalization of human fibroblasts and mammary epithelial cells transfected with SV40 large T-antigen. Exp Cell Res 209:45–52

Shay JW, Wright WE, Brasiskyte D, Van der Haegen BA (1993b) E6 of human papillomavirus type 16 can overcome the M1 stage of immortalization in human mammary epithelial cells but not in human fibroblasts. Oncogene 8:1407–1413

Shay JW, Tomlinson G, Piatyszek MA, Gollahon LS (1995) Spontaneous in vitro immortalization of breast epithelial cells from a patient with Li-Fraumeni syndrome. Mol Cell Biol 15:425–432

Slamon DJ, Godolphin W, Jones LA, Holt JA, Wong SG, Keith DE, Levin WJ, Stuart SG, Udove J, Ullrich A, Press MF (1989) Studies of the HER-2/neu proto-oncogene in human breast and ovarian cancer. Science 244:707–712

Smith HS (1991) In vitro models in human breast cancer. In: Harris JR, Hellman S, Henderson IC, D. W. Kinne DW (eds) Breast Diseases. JB Lippincott, Philadelphia, pp 181–189

Smith HS, Wolman SR, Dairkee SH, Hancock MC, Lippman M, Leff A, Hackett AJ (1987) Immortalization in culture: occurrence at a late stage in the progression of breast cancer. J Natl Cancer Inst 78:611–615

Soule HD, McGrath CM (1986) A simplified method for passage and long-term growth of human mammary epithelial cells. In Vitro Cell Dev Biol 22:6–12

Soule HD, Maloney TM, Wolman SR, Peterson WD Jr, Brenz R, McGrath CM, Russo J, Pauley RJ, Jones RF, Brooks SC (1990) Isolation and characterization of a spontaneously immortalized human breast epithelial cell line, MCF-10. Cancer Res 50:6075–6086

Stampfer M, Hallowes RC, Hackett AJ (1980) Growth of normal human mammary cells in culture. In Vitro 16:415–425

Stampfer MR, Bartley JC (1985) Induction of transformation and continuous cell lines from normal human mammary epithelial cells after exposure to benzo[a]pyrene. Proc Natl Acad Sci USA 82:2394–2398

Stampfer MR, Yaswen P (1992) Factors influencing growth and differentiation of normal and transformed human mammary epithelial cells in culture. In: Milo GE, Casto BC, Shuler CF (eds) Transformation of human epithelial cells: molecular and oncogenetic mechanisms. CRC Press, Boca Raton, pp 117–140

Stampfer MR, Pan CH, Hosoda J, Bartholomew J, Mendelsohn J, Yaswen P (1993a) Blockage of EGF receptor signal transduction causes reversible arrest of normal and immortal human mammary epithelial cells with synchronous re-entry into the cell cycle. Exp Cell Res 208:175–188

Stampfer MR, Yaswen P, Alhadeff M, Hosoda J (1993b) TGF beta induction of extracellular matrix associated proteins in normal and transformed human mammary epithelial cells in culture is independent of growth effects. J Cell Physiol 155:210–221

Stampfer MR, Bodnar A, Garbe J, Wong M, Pan A, Villeponteau B, Yaswen P (1997) Gradual phenotypic conversion associated with immortalization of cultured human mammary epithelial cells. Mol Biol Cell 8:2391–2405

Swisshelm K, Leonard M, Sager R (1992) Preferential chromosome loss in human papilloma virus DNA-immortalized mammary epithelial cells. Genes Chromosomes Cancer 5:219–226

Taylor-Papadimitriou J, Stampfer M, Bartek J, Lewis A, Boshell M, Lane EB, Leigh IM (1989) Keratin expression in human mammary epithelial cells cultured from normal and malignant tissue: relation to in vivo phenotypes and influence of medium. J Cell Sci 94:403–413

Teixeira MR, Pandis N, Bardi G, Andersen JA, Heim S (1996) Karyotypic comparisons of multiple tumorous and macroscopically normal surrounding tissue samples from patients with breast cancer. Cancer Res 56:855–9

Veldman T, Vignon C, Schrock E, Rowley JD, Ried T (1997) Hidden chromosome abnormalities in haematological malignancies detected by multicolour spectral karyotyping. Nat Genet 15:406–410

Vogel VG (1991) High-risk populations as targets for breast cancer prevention trials. Prev Med 20:86–100

Walen KH, Stampfer MR (1989) Chromosome analyses of human mammary epithelial cells at stages of chemical-induced transformation progression to immortality. Cancer Genet Cytogenet 37:249–261

Wazer DE, Chu Q, Liu XL, Gao Q, Safaii H, Band V (1994) Loss of p53 protein during radiation transformation of primary human mammary epithelial cells. Mol Cell Biol 14:2468–2478

Wazer DE, Liu XL, Chu Q, Gao Q, Band V (1995) Immortalization of distinct human mammary epithelial cell types by human papilloma virus 16 E6 or E7. Proc Natl Acad Sci USA 92:3687–3691

Weinberg RA (1996) E2F and cell proliferation: a world turned upside down. Cell 85:457–459

Weinberg RA (1998) Telomeres. Bumps on the road to immortality. Nature 396:23–24

Wolman SR, Mohamed AN, Heppner GH, Soule HD (1994) Chromosomal markers of immortalization in human breast epithelium. Genes Chromosomes Cancer 10:59–65

Xu C, Meikrantz W, Schlegel R, Sager R (1995) The human papilloma virus 16E6 gene sensitizes human mammary epithelial cells to apoptosis induced by DNA damage. Proc Natl Acad Sci USA 92:7829–7833

Yashima K, Milchgrub S, Gollahon LS, Maitra A, Saboorian MH, Shay JW, Gazdar AF (1998) Telomerase enzyme activity and RNA expression during the multistage pathogenesis of breast carcinoma. Clin Cancer Res 4:229–234

Zajchowski DA, Band V, Trask DK, Kling D, Connolly JL, Sager R (1990) Suppression of tumor-forming ability and related traits in MCF-7 human breast cancer cells by fusion with immortal mammary epithelial cells. Proc Natl Acad Sci USA 87:2314–2318

Zhai YF, Beittenmiller H, Wang B, Gould MN, Oakley C, Esselman WJ, Welsch CW (1993) Increased expression of specific protein tyrosine phosphatases in human breast epithelial cells neoplastically transformed by the neu oncogene. Cancer Res 53:2272–2278

Telomeres and Cell Division Potential

K. Perrem[1] and R. R. Reddel[1]

1
Introduction

Telomeres are specialized structures which are located at the ends of all linear eukaryotic chromosomes and which, with few exceptions, consist of G-rich DNA repeat sequences (Blackburn 1991) and specific telomere binding proteins (Zhong et al. 1992; Cardenas et al. 1993; Shore 1994; Chong et al. 1995; Broccoli et al. 1997). All vertebrate chromosome terminal sequences consist of TTAGGG hexanucleotide repeats (reading 5'-3' centromere to telomere; Moyzis et al. 1988; Blackburn 1991) and these have been estimated to range in length between about 4 and 15 kb in normal human cells (reviewed in Harley 1991). Unique chromatin arrangements at the telomere are facilitated by interactions between binding proteins and telomeric repeats (Makarov et al. 1993; Tommerup et al. 1994; Lejnine et al. 1995). It was postulated many decades ago without any specific knowledge of their structure that telomeres functioned as a protective cap that prevented chromosomes from end degradation and terminal fusion (Mueller 1938; McClintock 1941). This was subsequently demonstrated in molecular studies using cloned telomeric sequences (Bourgain and Katinka 1991). It has also been shown that telomeres have a role in chromosome segregation during mitosis (Kirk et al. 1997).

In addition to these critically important biological functions, it has been proposed that the telomere has a central role in determining the number of times cells can divide. In this chapter we review the evidence for this proposal.

2
The Telomere Hypothesis of Senescence

Hayflick found that primary human cells were incapable of unlimited division cycles in vitro (Hayflick and Moorhead 1961; Hayflick 1965). When the cells eventually ceased proliferation they manifested morphological

---

[1] Cancer Research Unit, Children's Medical Research Institute, 214 Hawkesbury Road, Westmead, Sydney, New South Wales 2145, Australia.

Progress in Molecular and Subcellular Biology, Vol. 24
A. Macieira-Coelho
© Springer-Verlag Berlin Heidelberg 1999

and other changes that were interpreted as being indicative of cellular aging, and the cells were therefore described as senescent. The maximum number of cell divisions of normal cells in culture is often called the "Hayflick limit".

Olovnikov (1971) and Watson (1972) independently recognized that the conventional DNA replication apparatus fails to copy a short segment at one end of each linear DNA molecule. This became known as the "end replication problem" (Levy et al. 1992). Olovnikov and others proposed the telomere hypothesis of senescence, namely that the progressive telomeric shortening resulting from cell division eventually causes cellular senescence (Olovnikov 1971; Olovnikov 1973; Harley et al. 1992).

Evidence has subsequently been obtained that telomere shortening may also occur due to a process, possibly the action of a putative 5'-3' exonuclease, that creates a 100–200 bp single-stranded overhang at the end of the telomere after each cell division cycle (Wellinger et al. 1996; Makarov et al. 1997; Fig 1). Unrepaired double strand breaks in the telomere would also result in telomere shortening. Under some circumstances yeast telomeres undergo rapid shortening (Li and Lustig 1996), and there is also evidence that rapid telomeric shortening may sometimes occur in human cells (Murnane et al. 1994; K. Perrem et al., unpubl. data).

A corollary of the telomere hypothesis of senescence is that unlimited proliferative potential requires a telomere maintenance mechanism (TMM). This is supported by a substantial body of empirical data which will be discussed briefly below before returning to an appraisal of the evidence for the telomere hypothesis itself.

3
Telomere Maintenance

Telomeres may be maintained in some species by retrotransposition (reviewed in Biessmann and Mason 1997), or by a recombination mechanism (Pluta and Zakian 1989; Lundblad and Blackburn 1993; Roth et al. 1997). In most eukaryotic species, however, telomeres are normally maintained through the activity of an enzyme, telomere terminal transferase (telomerase), first identified in *Tetrahymena thermophila* (Greider and Blackburn 1985). Telomerase was shown to be a ribonucleoprotein reverse transcriptase complex that utilizes an RNA component as a template for the de novo synthesis and sequential addition of the G-rich repeats onto telomeres (Greider and Blackburn 1985). Subsequent studies resulted in the cloning and characterization of the RNA component (TER) of human (Feng et al. 1995) and mouse (Blasco et al. 1995) telomerase. Telomerase has more than one protein subunit and the first of these to be identified in mammals was a protein of unknown function now referred to as TEP1 (Harrington et al. 1997a; Nakayama et al. 1997). Identification of conserved reverse transcriptase domains in the telomerase catalytic subunits of yeast and ciliates

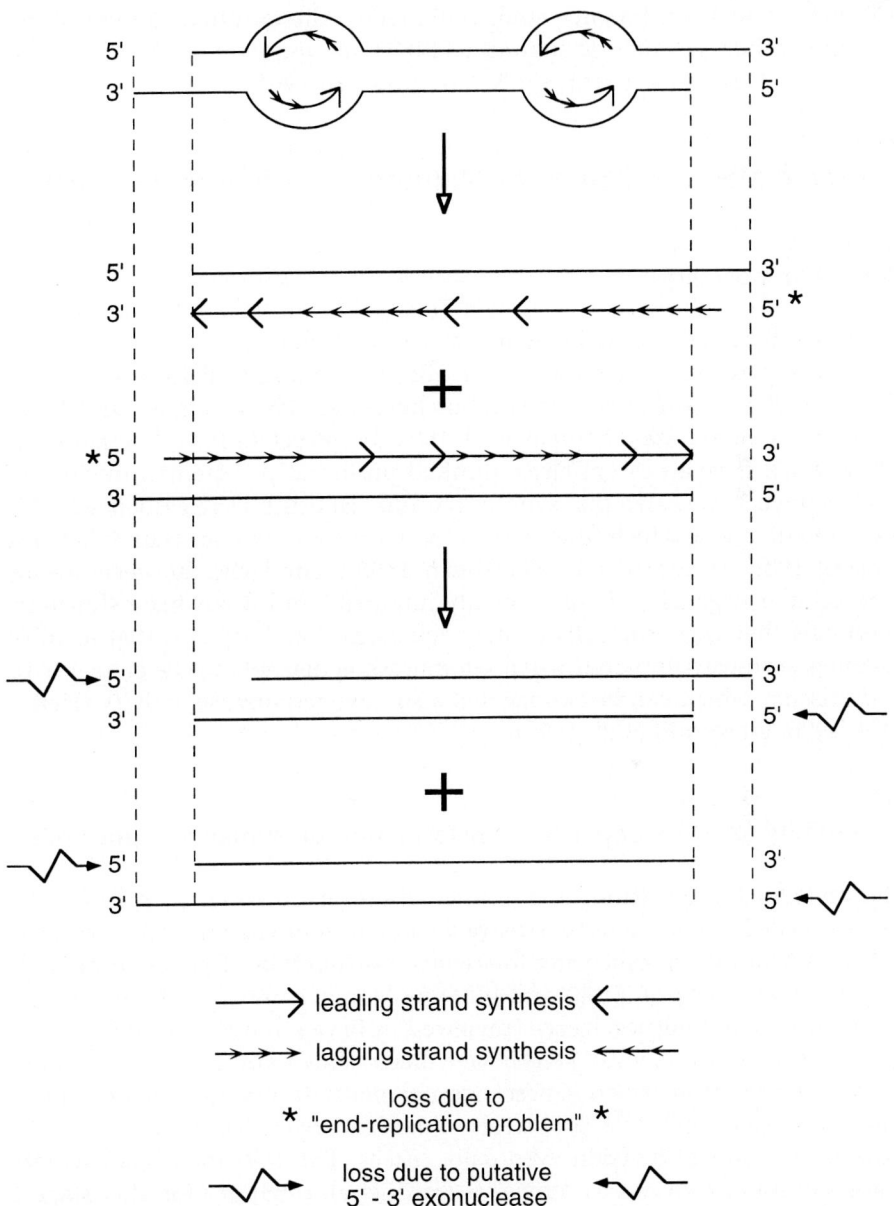

Fig. 1. Proposed mechanisms of telomeric DNA loss. DNA synthesis is RNA-primed, and the RNA primers are degraded after DNA synthesis is completed. Degradation of the RNA primer closest to the termini (indicated by the *asterisks*) generated by lagging-strand synthesis results in a short region that remains single-stranded. This is referred to as the "end-replication problem". Single-stranded overhangs are also generated and/or extended by a process that may involve a 5'-3' exonuclease. It is not clear whether all telomeres have this overhang (Makarov et al. 1997; Wright et al. 1997), but this process results in a shortened template for the next round of DNA synthesis. 5' indicates the CA-rich strand, and 3' the GT-rich strand

(Counter et al. 1997; Lingner et al. 1997) led to the eventual cloning of the human telomerase catalytic subunit hTERT (Harrington et al. 1997b; Kilian et al. 1997; Meyerson et al. 1997; Nakamura et al. 1997).

4
Telomere Maintenance and Unlimited Cell Division Potential

4.1
Normal Immortality

Telomere structure and maintenance has been studied in a variety of unicellular eukaryotes such as *Tetrahymena* (Blackburn and Gall 1978; Collins et al. 1995), *Euplotes aediculatus* (Klobutcher et al. 1981; Lingner and Cech 1996) and *S. cerevisiae* (Shampay et al. 1984; Wright et al. 1992; Lingner et al. 1997) which have an essentially unlimited proliferative potential in culture. Studies in yeast revealed that senescence and cell death were associated with mutation of a gene which functions in telomere maintenance (Lundblad and Szostak 1989; Lundblad and Blackburn 1993). Similarly, the germline of multicellular organisms is functionally immortal and it has been shown in mammals that germline cells contain telomerase activity and that fertility becomes seriously impaired when telomerase is disrupted (Lee et al. 1998). Plant tissues which can be propagated also have telomerase activity (Heller et al. 1996; Fitzgerald et al. 1996).

4.2
Immortality and Preneoplastic Transformation of Human Somatic Cells

The abnormal acquisition of unlimited division potential in mammalian somatic cells has been most extensively studied in in vitro models of human cell immortalization, especially following transduction of genes from DNA tumor viruses such as SV40 and other polyomaviruses, papillomaviruses, adenoviruses and herpesviruses (reviewed in Bryan and Reddel 1994). Cells which express these viral genes have been shown to continue dividing beyond the point at which senescence normally occurs and thus exhibit a "life span extension". This proliferation past the usual senescence barrier is temporary, and cell division eventually ceases. The cells then reach a state that is commonly known as "crisis" (Girardi et al. 1965) or mortality stage 2 (M2; Wright et al. 1989). Life-span extension in fibroblasts may also occur through loss of function of the p53 tumor suppressor gene or its major downstream effector p21 (Bond et al. 1994; Rogan et al. 1995; Brown et al. 1997), and in breast epithelial cells through loss of function of the p16[INK4] tumor suppressor gene (Brenner et al. 1998; Foster et al. 1998; Huschtscha et al. 1998).

The relationship, if any, between senescence and crisis is currently not well understood; it is not clear to what extent crisis may be a delayed form of senescence (Rubelj and Pereira-Smith 1994) or an imbalance between cell division and cell death rates resulting in a reduction in cell number (Girardi et al. 1965). Nevertheless, the occurrence of *limited* life span extension clearly indicates that even when the "normal" senescence checkpoint is bypassed there are other growth arrest checkpoints that are still capable of preventing unlimited proliferation (reviewed in Reddel 1998).

Rare clonal outgrowths from populations of transformed cells in crisis may occur and this often results in an apparently unlimited proliferative potential, i.e. an immortal phenotype (Girardi et al. 1965). The link between immortalization and telomere maintenance was first made by demonstrating that extracts of the immortal cervical carcinoma cell line HeLa contained telomerase activity (Morin 1989). Many studies have subsequently shown that immortalization of human cells is usually associated with activation of telomerase: activity of this enzyme was found to be absent from pre-crisis cells and present following escape from crisis (Counter et al. 1992; 1994; Kim et al. 1994; Klingelhutz et al. 1994; Bryan et al. 1995). Some immortalized cells did not exhibit telomerase activity but were shown to utilize a telomerase independent TMM (Murnane et al. 1994; Bryan et al. 1995), now known as alternative lengthening of telomeres (ALT; Bryan and Reddel 1997; Bryan et al. 1997; Reddel et al. 1997). No examples have been reported of immortalized human cell lines which do not utilize either telomerase or an ALT TMM. This is consistent with the idea that crisis is triggered by shortened telomeres and that escape from crisis requires acquisition of a TMM.

5
Telomere Maintenance and Escape from Senescence

5.1
Telomere Maintenance Is Not Required for Escape from Senescence

As predicted by the telomere hypothesis of senescence, it has been found in a number of studies that telomeres do shorten progressively with proliferation of cells in vitro (Harley et al. 1990; Allsopp et al. 1992; Counter et al. 1992). It is still not entirely clear whether this shortening is responsible for senescence. Cells which escape from senescence temporarily due to viral transformation (Counter et al. 1992; Counter et al. 1994; Klingelhutz et al. 1994) or due to loss of p53 (Rogan et al. 1995) or p21 (Brown et al. 1997) function, continue to undergo telomere shortening during the period of life span extension. Thus, telomere maintenance is not *necessary* for escape from senescence. Two ways of interpreting this are either that the telomere lengths in senescent cells do not pose a barrier to further cell division, or alternatively that telomere shortening does normally signal senescence but the signal can be overridden through the process of viral transformation.

5.2
Telomere Maintenance Is Sufficient for Escape from Senescence

In support of the interpretation that telomere shortening does normally trigger senescence, it has recently been found that expression of telomerase in some types of normal cells is *sufficient* for escape from senescence. Transient expression of telomerase catalytic subunit hTERT cDNA in telomerase negative cells was first shown to reconstitute telomerase activity (Weinrich et al. 1997; Counter et al. 1998; Wen et al. 1998). This suggested that the other components of telomerase were present and that the level of hTERT is the crucial factor in telomerase regulation. Control of hTERT expression appears to be by transcription and possibly also by alternative splicing (Kilian et al. 1997). It was then shown that stable expression of hTERT in normal human fibroblasts and some types of epithelial cells resulted in bypass of senescence (Bodnar et al. 1998; Vaziri and Benchimol 1998). In some types of epithelial cells, however, functional disruption of the retinoblastoma gene product pRb or of p16INK4 was also required (Kiyono et al. 1998). At the time of the initial reports, the amount of life span extension was at least 20–30 population doublings (PDs), but since then the cells have continued to proliferate for at least 70 PDs beyond their Hayflick limit, retain normal cell cycle checkpoint controls, and have accumulated remarkably few karyotypic changes (H. Vaziri, pers. comm.). Although it is not yet known whether the cells are immortalized, the highly significant extension of in vitro life span achieved in these experiments adds considerable weight to the telomere hypothesis of senescence.

There remain a number of caveats, however. It is at least possible that expression of hTERT has effects other than telomere maintenance and that these are responsible for the life span extension. It should be noted that expression of the adenovirus 12 E1B 54K protein (which is not known to have any effects on telomere maintenance) in normal human cells also resulted in life span extension of more than 100 PDs in some cases, and when these cells eventually ceased dividing terminal restriction fragment (TRF) analyses revealed insufficient telomere shortening to account for the proliferation arrest (Gallimore et al. 1997). Further, normal human oral keratinocytes senesced in culture without undergoing telomeric shortening (Kang et al. 1998). On the basis of these and other considerations it has been argued elsewhere that there may be more than one "mitotic clock" that can trigger senescence (Reddel 1998).

Another caveat is that the PD level at which the control cells are observed to become senescent in culture could possibly be well below their true proliferative potential. In the case of epithelial cells, conditions which do not completely prevent terminal differentiation can result in an artificially low proliferative potential. Further, there are a number of stimuli, including various forms of macromolecular damage other than telomere shortening, which can result in a senescence-like state sometimes referred to as prema-

Table 1. Stimuli that result in premature senescence, a senescence-like state at a population doubling level below the Hayflick limit

Reported stimuli	Ref.
Gamma-irradiation	Di Leonardo et al. (1994)
Oxidative stress	Chen and Ames (1994); Von Zglinicki et al. (1995); Chen et al. (1995); Bladier et al. (1997)
Activated Ha-ras oncogene	Hicks et al. (1991); Serrano et al. (1997); Lin et al. (1998)
Cross-linked DNA	Weeda et al. (1997)
DNA breaks	Robles and Adami (1998)
Ceramide	Venable et al. (1995)
Phosphatidylinositol-3-kinase inhibitor	Tresini et al. (1998)
DNA demethylating agents	Holliday (1986); Fairweather et al. (1987)
Histone deacetylase inhibitors	Ogryzko et al. (1996); Xiao et al. (1997)

ture senescence (Table 1) because it occurs prior to the Hayflick limit. It is very likely that optimal in vitro growth conditions have not yet been achieved for all cell types, in which case the Hayflick limit itself could be a form of premature senescence induced by cell culture-induced stress or damage. Telomeres appear to be particularly vulnerable to DNA damage (Von Zglinicki et al. 1995; Petersen et al. 1998; Sitte et al. 1998), so the effect of expressing telomerase in cultured normal cells could conceivably be to protect cells against this putative cell culture-induced damage. If this is correct, the possibility that there is a "true" Hayflick limit that is determined by something other than telomeric shortening cannot be ruled out yet. It will therefore be very interesting to determine whether the telomerase-expressing cells will eventually undergo senescence despite well-maintained telomeres.

6
Correlations Among In Vivo Age, Telomere Length, and Proliferative Potential

6.1
Correlation Between Telomere Shortening In Vivo and Age

To answer the question of whether currently available cell culture conditions cause abnormally rapid shortening of the telomeres of normal cells in vitro, it will be necessary to determine the rate at which telomeres shorten per PD under normal conditions in vivo. Technical limitations have made it difficult to obtain such data so far, but it has been possible to examine mean telomere length in various tissues as a function of age. It has been shown that the telomere length of human proliferative somatic cells (from tissues including colonic mucosa, blood and bone marrow) is inversely correlated with age (Harley et al. 1990; Hastie et al. 1990; Lindsey et al. 1991; Vaziri et al. 1993; Slagboom et al. 1994; Vaziri et al. 1994; Iwama et al. 1998). The rate of loss has been estimated to be 31–84 bp/year (Hastie et al. 1990; Vaziri et al. 1993; Slagboom et al. 1994; Iwama et al. 1998). A recent study showed that the rate of telomere loss was most rapid in infants up to the age of 4 years (>1 kb/year in peripheral blood lymphocytes) (Frenck et al. 1998), but greater numbers of individuals will need to be examined to confirm this. In contrast to these proliferative tissues, no decrease in telomere length of supposedly non-proliferative brain tissue was seen with age (Allsopp et al. 1995).

6.2
Lack of Correlation Between Age and Proliferative Potential

If telomere length determines proliferative capacity, and telomere length is inversely proportional to age, it would be expected that proliferative capacity would be inversely proportional to chronological in vivo age. A correlation was observed in several studies (Martin et al. 1970; Schneider and Mitsui 1976; Allsopp et al. 1992; Slagboom et al. 1994), but it was not strong and a large study of healthy individuals has recently found no correlation between age and the proliferative capacity of 124 fibroblast cell strains (Cristofalo et al. 1998). It would be interesting to know whether there was any correlation in these individuals between telomere length and proliferative capacity, as has been seen previously in a less extensive study (Allsopp et al. 1992).

6.3
Expression of Telomerase in Normal Tissues

One factor which potentially may complicate any relationships among age, telomere length, and cellular proliferative capacity is the ability of some normal cells to express telomerase activity (Table 2). Some evidence has indicated that the amount of telomerase expressed in normal hematopoietic cells is insufficient to prevent telomere shortening (Broccoli et al. 1995; Engelhardt et al. 1997). Recent data suggest that lymphocytes in germinal centres may express telomerase at levels sufficient to lengthen telomeres (reviewed in Weng et al. 1998), which may enable clonal expansion. It is thus possible that expression of telomerase is a hallmark of proliferating cells (Belair et al. 1997; Greider 1998), and that the relationship between age and proliferative capacity is modulated by the action of telomerase. However, it is unlikely that this accounts for the lack of correlation between age and proliferative

Table 2. Examples of normal human cells and tissues with telomerase activity[a]

Cell Type	Ref.
Hematopoietic Cells	Broccoli et al. (1995); Counter et al. (1995); Hiyama et al. (1995)
Epidermal Keratinocytes	Härle-Bachor and Boukamp (1996); Yasumoto et al. (1996); Taylor et al. (1996)
Hair Follicle Cells	Ramirez et al. (1997)
Intestine	Hiyama et al. (1996)
Uroepithelial Cells	Belair et al. (1997)
Endothelial Cells	Hsiao et al. (1997)
Liver	Burger et al. (1997)
Endometrium	Kyo et al. (1997); Saito et al. (1997); Brien et al. (1997)

[a] Data obtained from in vitro activity assays. In situ hybridization analysis has shown hTERT expression in a wide range of normal human tissues (Kolquist et al. 1998).

capacity of fibroblasts (Cristofalo et al. 1998), because several studies have reported absence of detectable telomerase activity in this cell type (e.g., Bryan et al. 1995). Alternative explanations for the lack of correlation may include heritable differences in telomere lengths (Slagboom et al. 1994), or it is possible, although somewhat speculative, that an ALT TMM is active in fibroblasts.

7
Evidence That Short Telomeres May Limit Proliferation In Vivo

The contribution, if any, of senescence to in vivo aging is still uncertain; it is not known whether senescent cells accumulate in aging tissues and thereby contribute to a decline in their functional capacity. This lack of information is at least partly due to the dearth of markers with which to identify senescent cells. One useful marker is senescence-associated beta-galactosidase activity, and an age-dependent increase in this marker was found in dermal fibroblasts and epidermal keratinocytes, supporting the concept that the accumulation of senescent cells may be part of the aging process (Dimri et al. 1995).

Regardless of whether senescent cells accumulate in vivo, an important question is whether shortened telomeres may be associated with or causally contribute to aging by resulting in decreased proliferative capacity. As will be discussed, studies of human tissues and of an animal model indicate that this may be the case.

7.1
Telomere Shortening in Human Vascular Endothelial Cells

The TRF length of human vascular endothelial cells decreases with increasing PDs in vitro, and the TRF length of endothelium from arteries subject to high hemodynamic stress declined as a function of age at a greater rate than endothelium from other arteries and from veins (Chang and Harley 1995). These data are consistent with the hypothesis that endothelial cells have a higher rate of turnover and therefore of proliferation, and consequently a faster rate of decline of TRF length with age, in vessels subject to high hemodynamic stress. It has not been possible yet, however, to demonstrate that endothelial cells in these vessels do replicate more frequently than such cells located elsewhere, nor to show directly whether the cells with shortened telomeres have reduced proliferative potential.

7.2
Telomere Shortening and Immunoexhaustion

In conditions where immunological function is deficient due to excessive cell turnover and replication, the telomere hypothesis would predict that telomeric shortening would accompany loss of proliferative capacity. This has been studied in the context of HIV-1-induced immune dysfunction, and the results have been complex. CD8(+) T cells were found to have shorter telomeres consistent with an increased turnover of these cells and exhaustion of the CD8(+) T-cell responses in HIV-1 infection. For CD4(+) T cells, telomere length analyses do not seem to provide evidence for replicative exhaustion, although the relationship here between telomere length and cell turnover remains controversial (reviewed in Wolthers and Miedema 1998).

7.3
Telomere Shortening and Cellular Proliferation Capacity
in the Telomerase-Negative Mouse

The most compelling evidence for a relationship between telomere length and proliferative capacity in vivo has come from mice rendered telomerase-negative by targeted disruption of the mTER gene (Blasco et al. 1997). Although cells from these mice are readily able to become immortalized in vitro, presumably due to an ALT mechanism, and are able to form tumors following transduction with an activated oncogene (Blasco et al. 1997), progressive telomeric shortening occurred with successive generations and was associated with decreased proliferative potential in the hematopoietic system and the testis (Lee et al. 1998). The presence of senescent cells in these or other tissues was not reported, but there was an increase in apoptosis (Lee et al. 1998). Some chromosome termini in mTER-negative cells lacked a detectable telomeric DNA sequence, and there was evidence for increased end-to-end fusion. These changes, which resemble those in cultured human cells in crisis rather than in senescence, may trigger apoptosis rather than senescence, at least in testicular and hematopoietic tissues. Although this result clearly shows a relationship between extreme telomere shortening and a decreased proliferative potential in vivo, caution must be exercised in drawing conclusions about normal aging where telomere shortening does not occur to this extent.

8
Perspectives

The telomere hypothesis of senescence is supported by the observation that telomeres shorten with proliferation of normal human cells in vitro and that expression of telomerase in otherwise normal cells is capable of increasing their proliferative capacity. There are a number of questions which remain to be answered, however, and confirmation of the hypothesis will require more detailed analysis of telomere lengths in individual cells which become senescent, and experimental induction of senescence by shortening of one or more telomeres in young cells. A signaling pathway which leads from shortened telomeres to induction of senescence needs to be elucidated.

It is not yet clear to what extent these in vitro observations reflect the situation in vivo. It is possible that in the normal in vivo environment the controlled actions of telomerase and maybe one or more alternative TMMs are usually capable of preventing telomere shortening from limiting proliferation. In this regard an important question which needs to be answered is why cells lose telomerase activity when cultured in vitro (Klingelhutz et al. 1996; Kang et al. 1998).

It has not yet been clearly demonstrated whether cellular senescence contributes to organismal aging, although cells with at least some of the characteristics of senescence have been identified in tissues in vivo. Nevertheless, extreme telomere shortening resulting from lack of telomerase activity for several generations has been shown in a mouse model to result in reduced in vivo proliferation. If it can be shown that telomere shortening-induced reduction in proliferative capacity contributes to aging in some tissues, it is at least possible that some aspects of aging may be able to be ameliorated by the development of treatments which act to restore telomere length.

Cells which undergo an abnormal extension of life span due to functional deficiency of genes such as p53 or the pRb pathway appear to undergo critical telomere shortening such that continued proliferation is dependent on activation of a TMM: telomerase or an ALT mechanism. The immortalization which thus ensues appears to be an important feature of the cancer phenotype, and inhibitors of the TMMs may therefore be useful for treating cancer.

Acknowledgements. The authors thank Lindy Hodgkin for assistance with the manuscript. Work in this laboratory was supported by the Carcinogenesis Fellowship of the New South Wales Cancer Council and project grants from the National Health and Medical Research Council of Australia.

References

Allsopp RC, Vaziri H, Patterson C, Goldstein S, Younglai EV, Futcher AB, Greider CW, Harley CB (1992) Telomere length predicts replicative capacity of human fibroblasts. Proc Natl Acad Sci USA 89:10114–10118

Allsopp RC, Chang E, Kashefi-Aazam M, Rogaev EI, Piatyszek MA, Shay JW, Harley CB (1995) Telomere shortening is associated with cell division in vitro and in vivo. Exp Cell Res 220:194–200

Belair CD, Yeager TR, Lopez PM, Reznikoff CA (1997) Telomerase activity: a biomarker of cell proliferation, not malignant transformation. Proc Natl Acad Sci USA 94:13677–13682

Biessmann H, Mason JM (1997) Telomere maintenance without telomerase. Chromosoma 106:63–69

Blackburn EH (1991) Structure and function of telomeres. Nature 350:569–573

Blackburn EH, Gall JG (1978) A tandemly repeated sequence at the termini of the extrachromosomal ribosomal RNA genes in *Tetrahymena*. J Mol Biol 120:33–53

Bladier C, Wolvetang EJ, Hutchinson P, De Haan JB, Kola I (1997) Response of a primary human fibroblast cell line to H_2O_2: Senescence-like growth arrest or apoptosis? Cell Growth Differ 8:589–598

Blasco MA, Funk W, Villeponteau B, Greider CW (1995) Functional characterization and developmental regulation of mouse telomerase RNA. Science 269:1267–1270

Blasco MA, Lee H-W, Hande MP, Samper E, Lansdorp PM, DePinho RA, Greider CW (1997) Telomere shortening and tumor formation by mouse cells lacking telomerase RNA. Cell 91:25–34

Bodnar AG, Ouellette M, Frolkis M, Holt SE, Chiu C-P, Morin GB, Harley CB, Shay JW, Lichtsteiner S, Wright WE (1998) Extension of life span by introduction of telomerase into normal human cells. Science 279:349–352

Bond JA, Wyllie FS, Wynford-Thomas D (1994) Escape from senescence in human diploid fibroblasts induced directly by mutant p53. Oncogene 9:1885–1889

Bourgain FM, Katinka MD (1991) Telomeres inhibit end to end fusion and enhance maintenance of linear DNA molecules injected into the *Paramecium primaurelia* macronucleus. Nucleic Acids Res 19:1541–1547

Brenner AJ, Stampfer MR, Aldaz CM (1998) Increased *p16* expression with first senescence arrest in human mammary epithelial cells and extended growth capacity with *p16* inactivation. Oncogene 17:199–205

Brien TP, Kallakury BVS, Lowry CV, Ambros RA, Muraca PJ, Malfetano JH, Ross JS (1997) Telomerase activity in benign endometrium and endometrial carcinoma. Cancer Res 57:2760–2764

Broccoli D, Young JW, de Lange T (1995) Telomerase activity in normal and malignant hematopoietic cells. Proc Natl Acad Sci USA 92:9082–9086

Broccoli D, Smogorzewska A, Chong L, de Lange T (1997) Human telomeres contain two distinct Myb-related proteins, TRF1 and TRF2. Nat Genet 17:231–235

Brown JP, Wei W, Sedivy JM (1997) Bypass of senescence after disruption of p21[CIP1/WAF1] gene in normal diploid human fibroblasts. Science 277:831–834

Bryan TM, Reddel RR (1994) SV40-induced immortalization of human cells. Crit Rev Oncog 5:331–357

Bryan TM, Reddel RR (1997) Telomere dynamics and telomerase activity in in vitro immortalised human cells. Eur J Cancer 33:767–773

Bryan TM, Englezou A, Gupta J, Bacchetti S, Reddel RR (1995) Telomere elongation in immortal human cells without detectable telomerase activity. EMBO J 14:4240–4248

Bryan TM, Englezou A, Dalla-Pozza L, Dunham MA, Reddel RR (1997) Evidence for an alternative mechanism for maintaining telomere length in human tumors and tumor-derived cell lines. Nat Med 3:1271–1274

Burger AM, Bibby MC, Double JA (1997) Telomerase activity in normal and malignant mammalian tissues: feasibility of telomerase as a target for cancer chemotherapy. Br J Cancer 75:516–522

Cardenas ME, Bianchi A, de Lange T (1993) A Xenopus egg factor with DNA–binding properties characteristic of terminus-specific telomeric proteins. Genes Dev 7:883–894

Chang E, Harley CB (1995) Telomere length and replicative aging in human vascular tissues. Proc Natl Acad Sci USA 92:11190–11194

Chen Q, Ames BN (1994) Senescence-like growth arrest induced by hydrogen peroxide in human diploid fibroblast F65 cells. Proc Natl Acad Sci USA 91:4130–4134

Chen Q, Fischer A, Reagan JD, Yan L-J, Ames BN (1995) Oxidative DNA damage and senescence of human diploid fibroblast cells. Proc Natl Acad Sci USA 92:4337–4341

Chong L, van Steensel B, Broccoli D, Erdjument-Bromage H, Hanish J, Tempst P, de Lange T (1995) A human telomeric protein. Science 270:1663–1667

Collins K, Kobayashi R, Greider CW (1995) Purification of Tetrahymena telomerase and cloning of genes encoding the two protein components of the enzyme. Cell 81:677–686

Counter CM, Avilion AA, LeFeuvre CE, Stewart NG, Greider CW, Harley CB, Bacchetti S (1992) Telomere shortening associated with chromosome instability is arrested in immortal cells which express telomerase activity. EMBO J 11:1921–1929

Counter CM, Botelho FM, Wang P, Harley CB, Bacchetti S (1994) Stabilization of short telomeres and telomerase activity accompany immortalization of Epstein-Barr virus-transformed human B lymphocytes. J Virol 68:3410–3414

Counter CM, Gupta J, Harley CB, Leber B, Bacchetti S (1995) Telomerase activity in normal leukocytes and in hematologic malignancies. Blood 85:2315–2320

Counter CM, Meyerson M, Eaton EN, Weinberg RA (1997) The catalytic subunit of yeast telomerase. Proc Natl Acad Sci USA 94:9202–9207

Counter CM, Meyerson M, Eaton EN, Ellisen LW, Dickinson Caddle S, Haber DA, Weinberg RA (1998) Telomerase activity is restored in human cells by ectopic expression of hTERT (hEST2), the catalytic subunit of telomerase. Oncogene 16:1217–1222

Cristofalo VJ, Allen RG, Pignolo RJ, Martin BG, Beck JC (1998) Relationship between donor age and the replicative life span of human cells in culture: a re-evaluation. Proc Natl Acad Sci USA 95:10614–10619

Di Leonardo A, Linke SP, Clarkin K, Wahl GM (1994) DNA damage triggers a prolonged p53-dependent G1 arrest and long-term induction of Cip1 in normal human fibroblasts. Genes Dev 8:2540–2551

Dimri GP, Lee XH, Basile G, Acosta M, Scott C, Roskelley C, Medrano EE, Linskens M, Rubelj I, Pereira-Smith O, Peacocke M, Campisi J (1995) A biomarker that identifies senescent human cells in culture and in aging skin in vivo. Proc Natl Acad Sci USA 92:9363–9367

Engelhardt M, Kumar R, Albanell J, Pettengell R, Han W, Moore MAS (1997) Telomerase regulation, cell cycle, and telomere stability in primitive hematopoietic cells. Blood 90:182–193

Fairweather DS, Fox M, Margison GP (1987) The in vitro life span of MRC-5 cells is shortened by 5-azacytidine-induced demethylation. Exp Cell Res 168:153–159

Feng J, Funk WD, Wang S-S, Weinrich SL, Avilion AA, Chiu C-P, Adams RR, Chang E, Allsopp RC, Yu JH, Le SY, West MD, Harley CB, Andrews WH, Greider CW, Villeponteau B (1995) The RNA component of human telomerase. Science 269:1236–1241

Fitzgerald MS, McKnight TD, Shippen DE (1996) Characterization and developmental patterns of telomerase expression in plants. Proc Natl Acad Sci USA 93:14422–14427

Foster SA, Wong DJ, Barrett MT, Galloway DA (1998) Inactivation of p16 in human mammary epithelial cells by CpG island methylation. Mol Cell Biol 18:1793–1801

Frenck RW Jr, Blackburn EH, Shannon KM (1998) The rate of telomere sequence loss in human leukocytes varies with age. Proc Natl Acad Sci USA 95:5607–5610

Gallimore PH, Lecane PS, Roberts S, Rookes SM, Grand RJA, Parkhill J (1997) Adenovirus type 12 early region 1B 54K protein significantly extends the life span of normal mammalian cells in culture. J Virol 71:6629–6640

Girardi AJ, Jensen FC, Koprowski H (1965) SV40-induced transformation of human diploid cells: crisis and recovery. J Cell Comp Physiol 65:69–84

Greider CW (1998) Telomerase activity, cell proliferation, and cancer. Proc Natl Acad Sci USA 95:90–92

Greider CW, Blackburn EH (1985) Identification of a specific telomere terminal transferase activity in *Tetrahymena* extracts. Cell 43:405–413

Härle-Bachor C, Boukamp P (1996) Telomerase activity in the regenerative basal layer of the epidermis in human skin and in immortal and carcinoma-derived skin keratinocytes. Proc Natl Acad Sci USA 93:6476–6481

Harley CB, Futcher AB, Greider CW (1990) Telomeres shorten during ageing of human fibroblasts. Nature 345:458–460

Harley CB (1991) Telomere loss: Mitotic clock or genetic time bomb? Mutat Res 256:271–282

Harley CB, Vaziri H, Counter CM, Allsopp RC (1992) The telomere hypothesis of cellular aging. Exp Gerontol 27:375–382

Harrington L, McPhail T, Mar V, Zhou W, Oulton R, Amgen EST Program, Bass MB, Arruda I, Robinson MO (1997a) A mammalian telomerase-associated protein. Science 275:973–977

Harrington L, Zhou W, McPhail T, Oulton R, Yeung DSK, Mar V, Bass MB, Robinson MO (1997b) Human telomerase contains evolutionarily conserved catalytic and structural subunits. Genes Dev 11:3109–3115

Hastie ND, Dempster M, Dunlop MG, Thompson AM, Green DK, Allshire RC (1990) Telomere reduction in human colorectal carcinoma and with ageing. Nature 346:866–868

Hayflick L (1965) The limited in vitro lifetime of human diploid cell strains. Exp Cell Res 37:614–636

Hayflick L, Moorhead PS (1961) The serial cultivation of human diploid cell strains. Exp Cell Res 25:585–621

Heller K, Kilian A, Piatyszek MA, Kleinhofs A (1996) Telomerase activity in plant extracts. Mol Gen Genet 252:342–345

Hicks GG, Egan SE, Greenberg AH, Mowat M (1991) Mutant p53 tumor suppressor alleles release *ras*-induced cell cycle growth arrest. Mol Cell Biol 11:1344–1352

Hiyama E, Hiyama K, Tatsumoto N, Kodama T, Shay JW, Yokoyama T (1996) Telomerase activity in human intestine. Int J Oncol 9:453–458

Hiyama K, Hirai Y, Kyoizumi S, Akiyama M, Hiyama E, Piatyszek MA, Shay JW, Ishioka S, Yamakido M (1995) Activation of telomerase in human lymphocytes and hematopoietic progenitor cells. J Immunol 155:3711–3715

Holliday R (1986) Strong effects of 5-azacytidine on the in vitro life span of human diploid fibroblasts. Exp Cell Res 166:543–552

Hsiao R, Sharma HW, Ramakrishnan S, Keith E, Narayanan R (1997) Telomerase activity in normal human endothelial cells. Anticancer Res 17:827–832

Huschtscha LI, Noble JR, Neumann AA, Moy EL, Barry P, Melki JR, Clark SJ, Reddel RR (1998) Loss of p16[INK4] expression by methylation is associated with life span extension of human mammary epithelial cells. Cancer Res 58:3508–3512

Iwama H, Ohyashiki K, Ohyashiki JH, Hayashi S, Yahata N, Ando K, Toyama K, Hoshika A, Takasaki M, Mori M, Shay JW (1998) Telomeric length and telomerase activity vary with age in peripheral blood cells obtained from normal individuals. Hum Genet 102:397–402

Kang MK, Guo W, Park N-H (1998) Replicative senescence of normal human oral keratinocytes is associated with the loss of telomerase activity without shortening of telomeres. Cell Growth Differ 9:85–95

Kilian A, Bowtell DDL, Abud HE, Hime GR, Venter DJ, Keese PK, Duncan EL, Reddel RR, Jefferson RA (1997) Isolation of a candidate human telomerase catalytic subunit gene, which reveals complex splicing patterns in different cell types. Hum Mol Genet 6:2011–2019

Kim NW, Piatyszek MA, Prowse KR, Harley CB, West MD, Ho PLC, Coviello GM, Wright WE, Weinrich SL, Shay JW (1994) Specific association of human telomerase activity with immortal cells and cancer. Science 266:2011–2015

Kirk KE, Harmon BP, Reichardt IK, Sedat JW, Blackburn EH (1997) Block in anaphase chromosome separation caused by a telomerase template mutation. Science 275:1478–1481

Kiyono T, Foster SA, Koop JI, McDougall JK, Galloway DA, Klingelhutz AJ (1998) Both Rb/p16^{INK4a} inactivation and telomerase activity are required to immortalize human epithelial cells. Nature 396:84–88

Klingelhutz AJ, Barber SA, Smith PP, Dyer K, McDougall JK (1994) Restoration of telomeres in human papillomavirus-immortalized human anogenital epithelial cells. Mol Cell Biol 14:961–969

Klingelhutz AJ, Foster SA, McDougall JK (1996) Telomerase activation by the E6 gene product of human papillomavirus type 16. Nature 380:79–82

Klobutcher LA, Swanton MT, Donini P, Prescott DM (1981) All gene-sized DNA molecules in four species of hypotrichs have the same terminal sequence and an unusual 3' terminus. Proc Natl Acad Sci USA 78:3015–3019

Kolquist KA, Ellisen LW, Counter CM, Meyerson M, Tan LK, Weinberg RA, Haber DA, Gerald WL (1998) Expression of *TERT* in early premalignant lesions and a subset of cells in normal tissues. Nat Genet 19:182–186

Kyo S, Takakura M, Kohama T, Inoue M (1997) Telomerase activity in human endometrium. Cancer Res 57:610–614

Lee H-W, Blasco MA, Gottlieb GJ, Horner JW II, Greider CW, DePinho RA (1998) Essential role of mouse telomerase in highly proliferative organs. Nature 392:569–574

Lejnine S, Makarov VL, Langmore JP (1995) Conserved nucleoprotein structure at the ends of vertebrate and invertebrate chromosomes. Proc Natl Acad Sci USA 92:2393–2397

Levy MZ, Allsopp RC, Futcher AB, Greider CW, Harley CB (1992) Telomere end-replication problem and cell aging. J Mol Biol 225:951–960

Li B, Lustig AJ (1996) A novel mechanism for telomere size control in *Saccharomyces cerevisiae*. Genes Dev 10:1310–1326

Lin AW, Barradas M, Stone JC, Van Aelst L, Serrano M, Lowe SW (1998) Premature senescence involving p53 and p16 is activated in response to constitutive MEK/MAPK mitogenic signaling. Genes Dev 12:3008–3019

Lindsey J, McGill NI, Lindsey LA, Green DK, Cooke HJ (1991) In vivo loss of telomeric repeats with age in humans. Mutat Res 256:45–48

Lingner J, Cech TR (1996) Purification of telomerase from *Euplotes aediculatus* : requirement of a primer 3' overhang. Proc Natl Acad Sci USA 93:10712–10717

Lingner J, Hughes TR, Shevchenko A, Mann M, Lundblad V, Cech TR (1997) Reverse transcriptase motifs in the catalytic subunit of telomerase. Science 276:561–567

Lundblad V, Blackburn EH (1993) An alternative pathway for yeast telomere maintenance rescues est1⁻ senescence. Cell 73:347–360

Lundblad V, Szostak JW (1989) A mutant with a defect in telomere elongation leads to senescence in yeast. Cell 57:633–643

Makarov VL, Lejnine S, Bedoyan J, Langmore JP (1993) Nucleosomal organization of telomere-specific chromatin in rat. Cell 73:775–787

Makarov VL, Hirose Y, Langmore JP (1997) Long G tails at both ends of human chromosomes suggest a C strand degradation mechanism for telomere shortening. Cell 88:657–666

Martin GM, Sprague CA, Epstein CJ (1970) Replicative life span of cultivated human cells. Effect of donor's age, tissue, and genotype. Lab Invest 23:86–92

McClintock B (1941) The stability of broken ends of chromosomes in *Zea mays*. Genetics 26:234–282

Meyerson M, Counter CM, Eaton EN, Ellisen LW, Steiner P, Dickinson Caddle S, Ziaugra L, Beijersbergen RL, Davidoff MJ, Liu Q, Bacchetti S, Haber DA, Weinberg RA (1997) *hEST2*, the putative human telomerase catalytic subunit gene, is up-regulated in tumor cells and during immortalization. Cell 90:785–795

Morin GB (1989) The human telomere terminal transferase enzyme is a ribonucleoprotein that synthesizes TTAGGG repeats. Cell 59:521–529

Moyzis RK, Buckingham JM, Cram LS, Dani M, Deaven LL, Jones MD, Meyne J, Ratliff RL, Wu J-R (1988) A highly conserved repetitive DNA sequence, $(TTAGGG)_n$, present at the telomeres of human chromosomes. Proc Natl Acad Sci USA 85:6622–6626

Mueller HJ (1938) The remaking of chromosomes. Collect Net 13:181–198

Murnane JP, Sabatier L, Marder BA, Morgan WF (1994) Telomere dynamics in an immortal human cell line. EMBO J 13:4953–4962

Nakamura TM, Morin GB, Chapman KB, Weinrich SL, Andrews WH, Lingner J, Harley CB, Cech TR (1997) Telomerase catalytic subunit homologs from fission yeast and human. Science 277:955–959

Nakayama J-I, Saito M, Nakamura H, Matsuura A, Ishikawa F (1997) *TLP1*: a gene encoding a protein component of mammalian telomerase is a novel member of WD repeats family. Cell 88:1–20

Ogryzko VV, Hirai TH, Russanova VR, Barbie DA, Howard BH (1996) Human fibroblast commitment to a senescence-like state in response to histone deacetylase inhibitors is cell cycle dependent. Mol Cell Biol 16:5210–5218

Olovnikov AM (1971) Principle of marginotomy in template synthesis of polynucleotides. Dokl Akad Nauk, SSR 201:1496–1499

Olovnikov AM (1973) A theory of marginotomy. J Theor Biol 41:181–190

Petersen S, Saretzki G, Von Zglinicki T (1998) Preferential accumulation of single-stranded regions in telomeres of human fibroblasts. Exp Cell Res 239:152–160

Pluta AF, Zakian VA (1989) Recombination occurs during telomere formation in yeast. Nature 337:429–433

Ramirez RD, Wright WE, Shay JW, Taylor RS (1997) Telomerase activity concentrates in the mitotically active segments of human hair follicles. J Invest Dermatol 108:113–117

Reddel R (1998) A reassessment of the telomere hypothesis of senescence. BioEssays 20: 977–984.

Reddel RR, Bryan TM, Murnane JP (1997) Immortalized cells with no detectable telomerase activity. A review. Biochemistry (Moscow) 62:1254–1262

Robles SJ, Adami GR (1998) Agents that cause DNA double strand breaks lead to $p16^{INK4a}$ enrichment and the premature senescence of normal fibrolasts. Oncogene 16:1113–1123

Rogan EM, Bryan TM, Hukku B, Maclean K, Chang AC-M, Moy EL, Englezou A, Warneford SG, Dalla-Pozza L, Reddel RR (1995) Alterations in p53 and $p16^{INK4}$ expression and telomere length during spontaneous immortalization of Li-Fraumeni syndrome fibroblasts. Mol Cell Biol 15:4745–4753

Roth CW, Kobeski F, Walter MF, Biessmann H (1997) Chromosome end elongation by recombination in the mosquito *Anopheles gambiae*. Mol Cell Biol 17:5176–5183

Rubelj I, Pereira-Smith OM (1994) SV40-transformed human cells in crisis exhibit changes that occur in normal cellular senescence. Exp Cell Res 211:82–89

Saito T, Schneider A, Martel N, Mizumoto H, Bulgay-Moerschel M, Kudo R, Nakazawa H (1997) Proliferation-associated regulation of telomerase activity in human endometrium and its potential implication in early cancer diagnosis. Biochem Biophys Res Commun 231:610–614

Schneider EL, Mitsui Y (1976) The relationship between in vitro cellular aging and in vivo human age. Proc Natl Acad Sci USA 73:3584–3588

Serrano M, Lin AW, McCurrach ME, Beach D, Lowe SW (1997) Oncogenic *ras* provokes premature cell senescence associated with accumulation of p53 and $p16^{INK4a}$. Cell 88:593–602

Shampay J, Szostak JW, Blackburn EH (1984) DNA sequences of telomeres maintained in yeast. Nature 310:154–157

Shore D (1994) RAP1: a protean regulator in yeast. Trends Genet 10:408–412

Sitte N, Saretzki G, Von Zglinicki T (1998) Accelerated telomere shortening in flbroblasts after extended periods of confluency. Free Radic Biol Med 24:885–893

Slagboom PE, Droog S, Boomsma DI (1994) Genetic determination of telomere size in humans: a twin study of three age groups. Am J Hum Genet 55:876–882

Taylor RS, Ramirez RD, Ogoshi M, Chaffins M, Piatyszek MA, Shay JW (1996) Detection of telomerase activity in malignant and nonmalignant skin conditions. J Invest Dermatol 106:759–765

Tommerup H, Dousmanis A, de Lange T (1994) Unusual chromatin in human telomeres. Mol Cell Biol 14:5777–5785

Tresini M, Mawal-Dewan M, Cristofalo VJ, Sell C (1998) A phosphatidylinositol 3-kinase inhibitor induces a senescent-like growth arrest in human diploid fibroblasts. Cancer Res 58:1–4

Vaziri H, Schächter F, Uchida I, Wei L, Zhu X, Effros R, Cohen D, Harley CB (1993) Loss of telomeric DNA during aging of normal and trisomy 21 human lymphocytes. Am J Hum Genet 52:661–667

Vaziri H, Dragowska W, Allsopp RC, Thomas TE, Harley CB, Lansdorp PM (1994) Evidence for a mitotic clock in human hematopoietic stem cells: loss of telomeric DNA with age. Proc Natl Acad Sci USA 91:9857–9860

Vaziri H, Benchimol S (1998) Reconstitution of telomerase activity in normal human cells leads to elongation of telomeres and extended replicative life span. Curr Biol 8:279–282

Venable ME, Lee JY, Smyth MJ, Bielawska A, Obeid LM (1995) Role of ceramide in cellular senescence. J Biol Chem 270:30701–30708

Von Zglinicki T, Saretzki G, Döcke W, Lotze C (1995) Mild hyperoxia shortens telomeres and inhibits proliferation of fibroblasts: a model for senescence? Exp Cell Res 220:186–193

Watson JD (1972) Origin of concatemeric T7 DNA. Nat New Biol 239:197–201

Weeda G, Donker I, De Wit J, Morreau H, Janssens R, Vissers CJ, Nigg A, Van Steeg H, Bootsma D, Hoeijmakers JHJ (1997) Disruption of mouse ERCC1 results in a novel repair syndrome with growth failure, nuclear abnormalities and senescence. Curr Biol 7:427–439

Weinrich SL, Pruzan R, Ma L, Ouellette M, Tesmer VM, Holt SE, Bodnar AG, Lichtsteiner S, Kim NW, Trager JB, Taylor RD, Carlos R, Andrews WH, Wright WE, Shay JW, Harley CB, Morin GB (1997) Reconstitution of human telomerase with the template RNA component hTR and the catalytic protein subunit hTRT. Nat Genet 17:498–502

Wellinger RJ, Ethier K, Labrecque P, Zakian VA (1996) Evidence for a new step in telomere maintenance. Cell 85:423–433

Wen J, Cong Y-S, Bacchetti S (1998) Reconstitution of wild-type or mutant telomerase activity in telomerase negative immortal human cells. Hum Mol Genet 7:1137–1141

Weng N-P, Hathcock KS, Hodes RJ (1998) Regulation of telomere length and telomerase in T and B cells: a mechanism for maintaining replicative potential. Immunity 9:151–157

Wolthers KC, Miedema F (1998) Telomeres and HIV-1 infection: in search of exhaustion. Trends Microbiol 6:144–147

Wright JH, Gottschling DE, Zakian VA (1992) Saccharomyces telomeres assume a non-nucleosomal chromatin structure. Genes Dev 6:197–210

Wright WE, Pereira-Smith OM, Shay JW (1989) Reversible cellular senescence: implications for immortalization of normal human diploid fibroblasts. Mol Cell Biol 9:3088–3092

Wright WE, Tesmer VM, Huffman KE, Levene SD, Shay JW (1997) Normal human chromosomes have long G-rich telomeric overhangs at one end. Genes Dev 11:2801–2809

Xiao H, Hasegawa T, Miyaishi O, Ohkusu K, Isobe K (1997) Sodium butyrate induces NIH3T3 cells to senescence-like state and enhances promoter activity of p21[WAF/CIP1] in p53-independent manner. Biochem Biophys Res Commun 237:457–460

Yasumoto S, Kunimura C, Kikuchi K, Tahara H, Ohji H, Yamamoto H, Ide T, Utakoji T (1996) Telomerase activity in normal human epithelial cells. Oncogene 13:433–439

Zhong Z, Shiue L, Kaplan S, de Lange T (1992) A mammalian factor that binds telomeric TTAGGG repeats in vitro. Mol Cell Biol 12:4834–4843

Cellular Mortality and Immortalization: A Complex Interplay of Multiple Gene Functions

R. Wadhwa[1], S. C. Kaul[2] and Y. Mitsui[2]

1
Cellular Mortality –
Restricted Replicative Potential of Normal Cells

Since the pioneering work of Hayflick and Moorhead (1961) it has been generally accepted that normal somatic cells when cultured can undergo only a limited number (depending on the cell type) of divisions and reach an irreversibly growth arrested, but viable stage. The age of cells is determined by the number of times cells divide rather than the calendar time elapsed. The restricted replicative capacity of normal cells which confers them the mortal divisional phenotype is widely accepted as the most consistent manifestation of cellular aging. Relevance of in vitro life span of cells to in vivo aging is evidenced by

(1) the correlation of in vitro life span and the donor age
(2) correlation between in vitro life span with the average life expectancy of the species, and
(3) the reduced life span of cells from patients afflicted with premature aging syndromes (Smith and Pereira-Smith 1996; Kaul et al. 1998a).

Replicative senescence is often perceived as one of the following four scenarios. The first one invokes its random stochastic origin and defines it as an incapability of the biological system to cope with accumulated damage by not being able to repair and maintain its machinery as a function of time. The second supports its genetic roots and proposes it as an inevitable outcome of the other vital functions that contribute to survival and propagation of a biological system. The third views it as an integral outcome of the activity of tumor suppressor genes that impede the progression of cell cycle and safeguard normal cells against uncontrolled proliferation. The fourth views it as an outcome of the end replication problem of chromosomes that is evident as precise shortening of telomeres with each cell division and is referred to as a molecular clock.

---

[1] Chugai Research Institute for Molecular Medicine, 153–2 Nagai, Ibaraki 300–41, Japan
[2] National Institute of Bioscience and Human-Technology, AIST, 1–1 Higashi, Tsukuba, Ibaraki 305–8566, Japan.

Progress in Molecular and Subcellular Biology, Vol. 24
A. Macieira-Coelho
© Springer-Verlag Berlin Heidelberg 1999

Since cells can withdraw from the cell cycle and become nondividing in response to culture conditions e.g., lack of mitogens, DNA damage or terminal differentiation, a major problem has been to distinguish these from senescence-related growth arrest. Senescent cells acquire several characteristic phenotypes including increased size, a more flattened and irregular shape, decreased membrane fluidity, increased protein oxidation, decreased DNA methylation, telomere shortening, endogenous β-gal activity, defects in mitogenic signaling and, most consistently, an irreversible cell cycle arrest in the G1/S and possibly in the G2/M boundaries of the cell cycle (Sherwood et al. 1988; Dimri et al. 1995; Holliday 1995; Rattan 1996; Smith and Pereira-Smith1996; Gonos 1998). Therefore, senescence is a metabolically active stage which is unresponsive to mitogenic signals, in contrast to quiescence which is overcome by stimulation with mitogens. Senescence is markedly different from other forms of cell death including apoptosis. In fact, senescent human fibroblasts have been shown to be strikingly resistant to apoptosis relative to their pre-senescent counterparts (Wang et al. 1994; Wang 1995). Such a failure of senescent cells to undergo apoptosis may result in their accumulation and deleterious effects in aged tissues.

2
Cellular Immortality- Escape from Senescence

Cells can escape from the limits on replicative capacity and can acquire the capacity to divide indefinitely in culture as a result of multiple mechanisms involving genetic and epigenetic changes. Human cells very seldom, if ever, become spontaneously immortalized. Oncogenes of certain DNA tumor viruses, such as large T antigen of the SV40 virus, the combination of E1A and E1B genes of adenoviruses, and E6 and E7 genes of the papillomaviruses when introduced in normal cells can result in life span extension after which cells enter a stage referred to as crisis. Only a very small number (at frequencies ranging from 10^{-6} to 10^{-9}) of cells usually escape from crisis and get immortalized by secondary events which lead to activation of pathways resulting in telomere maintenance. Genetic analysis has shown that immortalization is recessive in cell hybrids between immortal and normal cells of limited life span. The pairwise cell hybrids between different tumor cell lines have demonstrated four complementation groups (A-D) for immortality in which SV-40 transformed immortal cell lines were predominantly found to be in the 'A' group (Pereira-Smith and Smith 1988). Reintroduction of a normal chromosome 6 or 6q into the SV40-transformed immortal cells has been shown to result in growth inhibition and reappearance of a senescent-like morphology (Sandhu et al. 1994; Banga et al. 1997). Other chromosomes (1, 4, 7) have been reported to have similar effects on immortal cells of the B-D complementation groups (Sugawara et al. 1990; Ning et al. 1991; Ogata et al. 1993; Nakabayashi et al. 1997). These studies support a model in which inactivation of a limited number of loci are responsible for immortalization and such loci might be expected to encode growth suppressors.

3
Genes Involved in Maintenance of Mortal or Immortal Phenotypes

Replicative senescence and escape from it can be viewed as an outcome of the complex interplay of many gene functions, such as maintenance and repair of genome, fidelity of genetic information transfer, turnover of macromolecules, stress protein synthesis, scavenging of free radicals and balance of tumor suppressor and oncogenic activities. Recent experimental data has provided evidence that the genetic changes that lead to inactivation of p53 and/or pRB pathways and the ones that result in activation of telomere maintenance mechanism(s) are required to overcome barriers to immortalization. Studies on chromosomal transfer, however, seem to suggest the involvement of some yet to be characterized genetic elements that can govern the execution of senescence or immortalization programs. This chapter is focused on the current understanding of some of the gene activities which appear to be well associated with senescent or immortal phenotypes.

3.1
p53 Family of Tumor Suppressors

p53 is a transcription factor and tumor suppressor which inhibits cell cycle progression or induces apoptosis in response to stress or DNA damage (Giaccia and Kastan 1998). Ablation of p53 function by a variety of mechanisms is the most common event in human cancer, occurring in over half of all tumors. Several studies have shown that loss of wild type p53 function is associated with escape from senescence (Bond et al. 1994; Wynford-Thomas 1996a, b). Late passage cells were seen to possess higher DNA binding and transcriptional activation activities of p53 although there were no significant changes in the protein levels. When microinjected with anti-p53 antibodies, senescent cells undergo division (Gire and Wynford-Thomas 1998). Upregulation of p53 following DNA damage has been widely reported, and therefore the trigger for increased p53 activity during senescence may also elicit telomere shortening. p53-mediated growth arrest is due in part to the activity of cyclin-cdk inhibitor, $p21^{WAF-1}$, a member of p21 family of proteins ($p21^{WAF-1}$, $p27^{Kip-1}$ and $p57^{Kip-2}$). It binds to cyclin D1 and by limiting its kinase activity, results in underphosphorylated forms of pRB which remain bound to E2F, causing cell cycle arrest. Upregulation of $p21^{WAF-1}$ in senescent fibroblast cultures and extended population doubling of normal diploid fibroblasts by disruption of $p21^{WAF-1}$ have supported its role in cellular senescence (Noda et al. 1994; Brown et al. 1997). Other studies, however, have suggested that $p21^{WAF-1}$ is not a critical mediator in the tumor suppression function of p53. Mice lacking p21^{WAF-1} did not have a propensity for developing tumors (Brugarolas et al. 1995; Deng et al. 1995). Furthermore, p21^{WAF-1} upregulation by mitogenic and growth stimulatory signals has suggested that its physiological function is not limited to execution of cell cycle arrest programs. Recently, cell cycle arrest induced by p19^{ARF} (alternative reading frame coded by the INK4 locus) was shown to be mediated by wild type p53 (Chin et al. 1998; Sherr 1998) and p21^{WAF-1}. p19^{ARF} was shown to interact with p53 and mdm2, a negative regulator of p53 (Kubbutat et al. 1997; Zhang et al. 1998).

Three other members, p73a and p73b (Jost et al. 1997; Kaghad et al. 1997), p51A and p51B (Osada et al. 1998) and p40 (Trink et al. 1998) of the p53 tumor suppressor gene family have recently been cloned and shown to have functional analogy to p53 in mediating cell cycle arrest and apoptosis. p73, however, was not upregulated in response to DNA damage. The discovery of a p53 family of proteins evokes intriguing consequences, such as increased chances of occurrence of mutant p53-like proteins which may interfere with normal p53 function and on the other hand serve as back-up genes for p53-related normal functions. Therefore, the role of the growing family of p53 homologues in senescence-related growth arrest of cells, and escape from it to undergo immortalization, awaits further studies.

3.2
pRB and INK Family of Tumor Suppressors

In their hypophosphorylated active states, pRB and other pocket proteins, p107 and p130, retain cells in G1 phase by repressing the transcription of E2F responsive genes which are required for progression through the G1-S transition and S phase (Weinberg 1996). Hypophosphorylated pRB physically associates with promoter-bound E2F and induces repression by masking the E2F transactivation domain and recruiting histone deacetylase (HDAC), a modulator of core-histone-DNA interactions which is linked to repression of gene expression. Both pRB-E2F and pRB-HDAC interactions were shown to be disrupted by viral oncoproteins, which implies that these interactions play a critical role in tumor suppressor activity of pRB. Phosphorylation of pRB by active cyclin-cyclin dependent kinases (cdks) complexes results in release of E2F and leads to S-phase entry. Inhibitors of cyclin-cdks, p21^{WAF-1} (described in Sect. 3.1) and INK4 family members (p16^{INK4a}, p15^{INK4b}, p18^{INK4c} and p19^{INK4d}) which bind to cdk-4 and cdk-6 (Chin et al. 1998) lead to cell cycle arrest by impeding phosphorylation of the pRb family of proteins. Senescent cells, as expected, have increased levels of p16^{INK4a} and hypophosphorylated pRB (Seshadri and Campisi 1990; Hara et al. 1996). Furthermore, introduction of p16^{INK4a} induces cell cycle arrest and senescence in cells with functional pRb, and inactivation of p16^{INK4a} is reported to be involved in immortalization as an alternative to the loss of pRb (Yeager et al. 1995, 1998; Noble et al. 1996; Bartek et al. 1997).

The roles of p53 and pRb in cellular senescence have been elucidated by many independent studies (Kulju and Lehman 1995; Bond et al. 1996; Vaziri and Benchimol 1996; Serrano et al. 1997; von Zglinicki and Saretzki 1997; Gire and Wynford-Thomas 1998). The two pathways seem to converge to ensure a hypophosphorylated status, to impose a cell cycle checkpoint in normal cells. Inactivation of pRB and/or p53 function by SV40 T Ag is seen to cause life span extension of normal fibroblasts. RB+/- and p53-/- mice show tumors at a younger age than RB+/- or p53-/- alone. The DNA tumor viruses that inactivate both pRB and p53 can function more efficiently as transforming agents than those that encode oncoproteins which engage either pRB or p53, suggesting cooperation between pRB and p53. Although it is evident that the inactivation of both pRb and p53 is essential for cells to circumvent the normal proliferative constraints imposed by replicative senescence, there are many transformed cell lines that retain wild type p53 and/ pRb status (Whitaker et al. 1995), suggesting that there are alternative pathways to immortalization.

3.3
Ras/Raf/MEK/MAP Kinase Pathway

Oncoproteins Ras and Raf-1 transform established cell lines to neoplastic growth. However, in normal cells these genes induce premature senescence in the absence of cooperating oncogenes such as E1A, T-antigen, or myc. Recent work has shown that Ras mediated senescence is elicited by the Raf-1/MEK/MAP kinase signaling cascade (Lin et al. 1998; Zhu et al. 1998) which can induce p53, $p21^{WAF-1}$ and $p16^{INK4a}$ expressions. Oncogenic Ras and MEK also induce $p19^{ARF}$ expression, and ARF-/- mouse embryonic fibroblasts were defective in Ras induced growth arrest (Palmero et al. 1998), suggesting the involvement of $p19^{ARF}$ in Ras induced senescence. In another study, Ras did not arrest proliferation of $p21^{WAF-1}$ -/- mouse fibroblasts, suggesting a critical role for $p21^{WAF-1}$ in response to H-Ras or Raf-1. However, human fibroblasts were seen to undergo Raf-1 induced senescence even in the absence of p53 and $p21^{WAF-1}$, suggesting that the p53 pathway is not critically involved. On the other hand, $p16^{INK4a}$ expression was found to be sufficient for induction of the senescent phenotype in normal human fibroblasts. Consistent with the irreversible nature of senescence, $p16^{INK4a}$ levels remain elevated even after the inactivation of signals from Raf-1. By using p53-/- and $p16^{INK4a}$-/- cells, Lin et al. (1998) have concluded that Ras mediated MAPK activation can produce two precisely opposite outcomes, i.e., cell cycle arrest and forced mitogenesis, depending on the integrity of the senescence program controlled by p53 and p16. A number of interesting differences exist between the senescent phenotype elicited by Raf-1 and prolonged passage in culture. Senescent cells are typically arrested at the G1 phase of cell cycle. In contrast, only a portion of cells, rendered senescent by Raf-1, were arrested in G1 and the remainder was in G2/M. Ras induced senescent cells showed a flattened cell phenotype in contrast to the rounded refractile morphology of Raf-1 induced senescent cells. These data seem to suggest the involvement of other genes or pathways in the production of the characteristic senescent phenotype.

3.4
Telomerase

Telomeres, the repeat sequences (TTAGGG)n, that protect chromosome ends from damage and rearrangements are lost with each round of proliferation because of the nature of DNA replication. Germ line, hematopoietic and cancerous cells, however, maintain the length of their telomeres because they either express a ribonucleoprotein polymerase, telomerase, which adds new telomere repeats to the chromosomal ends (Harley et al. 1992; Counter 1996; Kim 1997), or have an alternative mechanism for lengthening of telomeres (ALT; Bryan et al. 1995; Reddel et al. 1997). A number of studies have shown that the telomere shortening may be a clock for replicative senescence. The cells with shortening telomeres eventually reach a state of genomic instability and die whereas variants with telomerase expression arise and form an immortal population of tumor cells. The expression of the telomerase catalytic subunit in presenescent human cells has been reported to extend their in vitro life span significantly (Bodnar et al. 1998). These results have implied that activation of telomerase can result in one-step immortalization. However, in the case of the common mouse lab strain, *Mus musculus*, the telomeres are about ten times longer than in human cells, yet the mouse cells undergo replicative senescence much sooner than human cells (Blasco et al. 1997). The mouse strain *Mus spretus* displays telomere length similar to that of humans, has telomerase activity in most normal tissues and undergoes telomere shortening with increasing population doubling of cells in vitro (Prowse and Greider 1995; Broccoli et al. 1996). Normal human hematopoietic cells also show telomere shortening in spite of the detectable amounts of telomerase. Blasco et al. (1997) have obtained mice which were deleted in their germ line for the RNA component of telomerase. Such mice exhibit normal development and their fibroblasts undergo immortalization in culture. Telomere shortening was observed in these cells suggesting that telomerase is essential for telomere maintenance, but is not required for the establishment of a cell line. A microcell fusion approach to introduce a single human chromosome to transformed cells to restore the senescence program has shown that introduction of some, but not all, chromosomes is associated with inactivation of telomerase. Many of the cell clones that senesce in culture were seen to have telomerase activity (Oshimura and Barrett 1997). Recently, Syrian hamster embryo cells were reported to undergo senescence in culture in spite of the presence of telomerase activity, and their telomeres did not decrease in length, implicating a senescence mechanism that does not involve telomere loss (Carman et al. 1998).

The activities of many other genes which have been characterized as associated with senescent or transformed cells are not described here (see the following reviews Wadhwa et al. 1991a, 1994; Barrett et al. 1994; Sasaki et al. 1994; Iatropoulos and Williams 1996; Chang 1997; Duncan and Reddel 1997; Haber 1997; von Zglinicki and Saretzki 1997; Berube et al. 1998; Kaul et al. 1998a). Other studies have suggested that changes in chromosome structure and function are involved in cellular senescence and aging (Guarente 1996,1997). Derepression of inactive genes in an age-dependent manner has been reported (Wareham et al. 1987) and inhibition of DNA methylation which is associated with transcriptional silencing is shown to decrease the number of population doublings prior to senescence (Holliday 1987). When human fibroblasts are treated with histone deacetylase inhibitors they arrest in a state that has all the hallmarks of normal senescence (Ogryzko et al. 1996). Yet many other studies have shown that proliferative properties of cells are influenced by proteins that are not involved, at least directly, in cell cycle regulation (Merrick et al. 1994; Bini et al. 1997).

Murine cells show frequent spontaneous immortalization in contrast to human cells that almost never spontaneously immortalize. Addition of DNA tumor viruses alone, mild mutagen treatments and even deregulated expression of cellular genes such as c-myc are enough in certain murine cells for immortalization. Crisis stage is either not observed or is very subtle in case of murine cells. These cells, due to their ability to immortalize spontaneously, were exploited to identify genes associated with cellular mortal or immortal phenotypes. Independent studies on spontaneously immortalized clones from different strains of mouse have shown that most of these clones had one or both mutant p53 alleles, in contrast to their parent cells which expressed wild type p53. The immortal clones which expressed wild type p53 were found to be defective in one or other elements of the pRb pathway.

A comparison of proteins from normal and immortal cells revealed that an actin-binding cytoskeletal protein, ezrin, was frequently upregulated in both mouse and human transformed cells, although it was insufficient for immortalization. The protein was shown to have a proproliferative function and was involved in pathways which negatively regulate contact inhibition (Kaul et al. 1996). Other cytoskeletal proteins including talin, vinculin, vimentin, α-actinin, tropomysin-1, lamin A and lamin C and/or localization of their protein products also change following immortalization (Kaneko et al. 1995). Since intermediary filaments may be involved in telomere binding, it has been speculated that alterations in these proteins may play a significant role in telomere maintenance during immortalization (Kaneko et al. 1995). Taken together, these studies have implicated a basic similarity in pathways that elicit senescence in human and mouse cells. Significantly, fusion hybrids of normal and immortal mouse cells yielded clones that senesced in culture, demonstrating that senescence in murine cells is dominant over immortalization (Wadhwa et al. 1991b) and therefore may involve pathways similar to the ones involved in human cells.

3.5
Mortalin

Protein screening between mouse cells with mortal and immortal phenotypes initially led to the identification of a protein of approximately 66 kDa
in cytoplasmic fractions of normal as well as mortal hybrid cells, suggesting
its association with the cellular mortal phenotype. Antigen isolated from
SDS-polyacrylamide gels was used to raise a polyclonal antibody that in turn
was used to clone the gene followed by subsequent functional analyses.
Sequence analyses of the clones isolated by immunoscreening assigned the
protein to the heat shock protein 70 (hsp70) family. Because of its identification as a protein associated with the cellular mortal phenotype, it was
named mortalin (Wadhwa et al. 1993a). Microinjection of an anti-mortalin
antibody to senescent mouse cells led to the transient stimulation of cell
division (Wadhwa et al. 1993a). Also, consistent with its antiproliferative
function, the transfection of mortalin cDNA isolated from normal cells to
NIH 3T3 cells was seen to induce the mortal phenotype (Wadhwa et al.
1993b). Subsequent immunofluorescence studies revealed that the protein
was not absent but localized to a different niche in immortal cells (Wadhwa
et al. 1993c). Whereas in normal cells the protein is widely distributed in the
cytoplasm (pancytosolic), the immortal cells harbor it in a perinuclear
locale (Fig. 1). Studies on human cells revealed that the protein is conserved
as a nearly 66 kDa mass. In normal human cells it exhibited pancytosolic
distribution whereas in human transformed cells, one of the four kinds of
staining patterns (i.e., granular staining with juxtanuclear concentration,
gradient of concentration from nuclear membrane to cell membrane, juxtanuclear arch, and fibrous staining with perinuclear concentration) were
observed (Wadhwa et al. 1995). The mouse pancytosolic and perinuclear
mortalin proteins are coded by two distinct genes, products of which differ

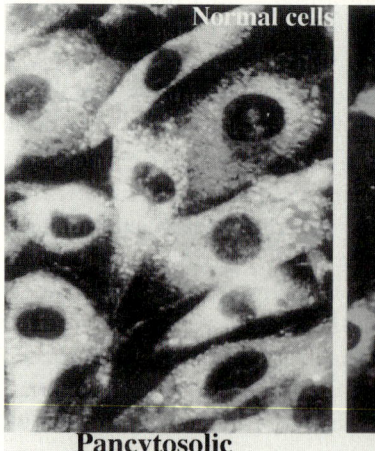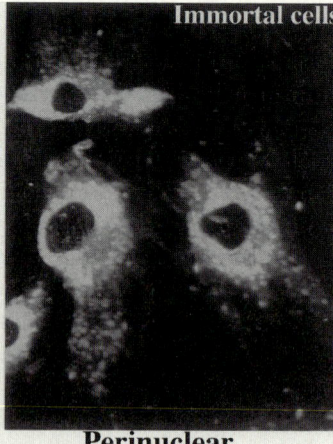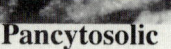

Fig. 1. Intracellular
distribution of
mortalins in normal
and immortal
mouse cells

by only two base pairs in the open reading frame (G1949A and G1959C), corresponding to two amino acid changes (V618M and R624G) near the carboxy-terminus of the protein in some strains of mouse and may involve other mechanisms such as interactions with other proteins (Wadhwa et al. 1993b, 1996). The latter seems to be more predominant in human cells in which cloning studies have not led to identification of four kinds of cDNAs encoding proteins which have differential cellular distributions. This model is also supported by heat shock induced change of the cellular distribution of the protein, albeit its heat uninducibility (Kaul et al. 1993).

Using biochemical fractionation, immunohistochemical and electron microscopy analysis, mortalin is indeed observed in different subcellular compartments in different human cell lines. Since the mortalin immunofluorescence patterns correlated with the assignment of the four complementation groups for immortalization (Pereira-Smith and Smith 1988; Wadhwa et al. 1995), it is suggested that there could be more than one event or pathways by which the normal distribution of mortalin can be perturbed, and each of these pathways can independently lead to immortalization. These observations imply that mortalin immunofluorescence can be exploited to define the genetic changes leading to immortalization. Support for this concept is also offered by studies on spontaneously immortalized mouse fibroblasts in which, irrespective of the strain, immortal cells were always seen to have perinuclear cellular distribution, supporting the concept that the loss of the pancytosolic form is essential or is tightly linked with the loss of cellular mortal phenotype. Fujii et al. (1995) have also shown that the distribution of mortalin in cells serves as a reliable marker of the divisional state of cells and varies with the mortal or immortal state in the same cells. Nakabayashi et al. (1997) have shown that the induction of cellular senescence in chromosome 7 radiation hybrids is accompanied by reversion of cellular distribution of mortalin from perinuclear to pancytosolic in the immortal human cell line, KMST-6.

In vivo studies by Northern analysis, RNA in situ hybridization, and immunohistochemistry in normal and tumor rat tissues have shown the highest level of mortalin expression in brain, heart, and skeletal muscle. Mortalin expression was significantly high in upper nondividing layers of skin, neurons and nerve fibers, cardiomyocytes and interstitial secretory tissue. Therefore, tissue- and cell-specific expression patterns of mortalin corresponds with the antiproliferative function in normal cells. Intriguingly, a deregulation of the expression was observed in rat brain tumors along with the detection of nonpancytosolic mortalin in rat glioma cell line, C6, indicating that the nonpancytosolic mortalin may have a role in tumorigenesis (Kaul et al. 1997). Studies were extended to human brain tumors that showed its progressive increase with malignancy, similar to a proliferation marker MIB-1 (Takano et al. 1997). Glioblastoma derived cell lines were seen to have non-pancytosolic mortalin immunofluorescence. An increase in mortalin expression was also observed in some postcrisis versus precrisis counterparts, suggesting that nonpancytosolic protein which is expressed in transformed cells may impart divisional advantage to the cells. The mot-2 overexpressing derivatives of NIH 3T3 were indeed seen to undergo malignant transformation (Kaul et al. 1998b). This phenotype is shown to be mediated, at least in part, by its interactions with tumor suppressor p53 (Wadhwa et al. 1998). Interestingly, the pancytosolic protein in normal cells was not found to colocalize with p53 nor the mot-1 cDNA affected transcriptional activity of p53 in reporter assays, suggesting that the senescence-inducing function of mot-1 may use a different pathway. Cytoplasmic sequestration of p53 by mot-2 was suggested as its possible mechanism for inactivation. These studies have demonstrated that interactions of mortalin with other cellular factors such as p53 may be instrumental for its association and role in cellular divisional phenotypes. Therefore, the eventual divisional phenotype of cells depends on various gene functions relating to the cellular background. Some other functions that have been assigned to mortalin include antigen processing (Domanico et al. 1993), response to glucose regulation (Mizzen et al. 1991), chaperoning of proteins and Ca^{2+} homeostasis (Webster et al. 1994; Massa et al. 1995). In addition, its similarity to Dnak suggests a function in DNA replication. Bruschi and Lindsay (1994) have reported that during TFEC-adduction, subunits of mitochondrial dehydrogenase multienzyme form tertiary complexes with mortalin/PBP74. Such studies can provide further guidelines to unravel functional mechanisms of mortalin in cellular mortality and immortalization.

4
Conclusions

Experimental data seem to favor the view that replicative senescence and escape from it is not controlled by a single mechanism. It involves the complex interplay of many gene functions by multiple pathways. Appreciation of the contributions of cell cycle regulatory genes such as p53, pRB and others such as telomerase to cellular senescence and immortalization, respectively, have made them the targets of future genetic therapies. At the same time, there are other genetic changes which, although poorly characterized, have a profound impact on different facets of senescence and immortalization programs. The combined interactions and integrity of genetic functions involving much more than just a handful of genes may determine the fate of cells to normal mortality or immortalization. Relevance of such interplay and their relationship continues to be a major challenge in the understanding of cellular mortality and immortalization.

References

Banga SS, Kim S, Hubbard K, Dasgupta T, Jha KK, Patsalis P, Hauptschein R, Gamberi B, Dalla-Favera R, Kraemer P, Ozer H L (1997) SEN6, a locus for SV40-mediated immortalization of human cells, maps to 6q26–27. Oncogene 14: 313–321

Barrett JC, Annab LA, Alcorta D, Preston G, Vojta P, Yin Y (1994) Cellular senescence and cancer. Cold Spring Harbor Symp Quant Biol 59: 411–418

Bartek J, Bartkova J, Lukas J (1997) The retinoblastoma protein pathway in cell cycle control and cancer. Exp Cell Res 237: 1–6

Berube NG, Smith JR, Pereira-Smith OM (1998) The genetics of cellular senescence. Am J Hum Genet 62: 1015–1019

Bini L, Magi B, Marzocchi B, Arcuri F, Tripodi S, Cintorino M, Sanchez JC, Frutiger S, Hughes G, Pallini V, Hochstrasser DF, Tosi P (1997) Protein expression profiles in human breast ductal carcinoma and histologically normal tissue. Electrophoresis 18: 2832–2841

Blasco MA, Lee HW, Rizen M, Hanahan D, DePinho R, Greider CW (1997) Mouse models for the study of telomerase. Ciba Found Symp 211: 160–170

Bodnar AG, Ouellette M, Frolkis M, Holt SE, Chiu CP, Morin GB, Harley CB, Shay J W, Lichtsteiner S, Wright W E (1998) Extension of life-span by introduction of telomerase into normal human cells. Science 279: 349–352

Bond JA, Wyllie FS, Wynford-Thomas D. (1994) Escape from senescence in human diploid fibroblasts induced directly by mutant p53. Oncogene 9: 1885–1889

Bond J, Haughton M, Blaydes J, Gire V, Wynford-Thomas D, Wyllie F (1996) Evidence that transcriptional activation by p53 plays a direct role in the induction of cellular senescence. Oncogene 13: 2097–2104

Broccoli D, Godley LA, Donehower LA, Varmus HE, de Lange T (1996) Telomerase activation in mouse mammary tumors: lack of detectable telomere shortening and evidence for regulation of telomerase RNA with cell proliferation. Mol Cell Biol 16: 3765–3772

Brown JP, Wei W, Sedivy JM (1997) Bypass of senescence after disruption of p21CIP1/WAF1 gene in normal diploid human fibroblasts. Science 277: 831–834

Brugarolas J, Chandrasekaran C, Gordon JI, Beach D, Jacks T, Hannon GJ (1995) Radiation-induced cell cycle arrest compromised by p21 deficiency. Nature 377: 552–557

Bruschi SA, Lindsay JG (1994) Mitochondrial stress protein actions during chemically induced renal proximal tubule cell death. Biochem Cell Biol 72: 663–667

Bryan TM, Englezou A, Gupta J, Bacchetti S, Reddel RR (1995) Telomere elongation in immortal human cells without detectable telomerase activity. EMBO J 14: 4240–4248

Carman TA, Afshari CA, Barrett JC (1998) Cellular senescence in telomerase-expressing Syrian hamster embryo cells. Exp Cell Res 244: 33–42

Chang ZF (1997) Regulatory mechanisms of replication growth limits in cellular senescence. J Formos Med Assoc 96: 784–791

Chin L, Pomerantz J, DePinho RA (1998) The INK4a/ARF tumor suppressor: one gene–two products–two pathways. Trends Biochem Sci 23: 291–296

Counter CM (1996) The roles of telomeres and telomerase in cell life span. Mutat Res 366: 45–63

Deng C, Zhang P, Harper JW, Elledge SJ, Leder P (1995) Mice lacking p21CIP1/WAF1 undergo normal development, but are defective in G1 checkpoint control. Cell 82: 675–84

Dimri GP, Lee X, Basile G, Acosta M, Scott G, Roskelley C, Medrano EE, Linskens, M, Rubelj I, Pereira-Smith OM, et al. (1995) A biomarker that identifies senescent human cells in culture and in aging skin in vivo. Proc Natl Acad Sci. USA 92: 9363–9967

Domanico SZ, DeNagel DC, Dahlseid JN, Green JM, Pierce SK (1993) Cloning of the gene encoding peptide-binding protein 74 shows that it is a new member of the heat shock protein 70 family. Mol Cell Biol 13: 3598–3610

Duncan EL, Reddel RR (1997) Genetic changes associated with immortalization. A review. Biochemistry (Mosc) 62: 1263–1274

Fujii M, Ide T, Wadhwa R, Tahara H, Kaul SC, Mitsui Y, Ogata T, Oishi M, Ayusawa D. (1995) Inhibitors of cGMP-dependent protein kinase block senescence induced by inactivation of T antigen in SV40-transformed immortal human fibroblasts. Oncogene 11: 627–634

Giaccia AJ, Kastan MB (1998) The complexity of p53 modulation: emerging patterns from divergent signals. Genes Dev 12: 2973–2983

Gire V, Wynford-Thomas D (1998) Reinitiation of DNA synthesis and cell division in senescent human fibroblasts by microinjection of anti-p53 antibodies. Mol Cell Biol 18: 1611–1621

Gonos ES (1998) Expression of the growth arrest specific genes in rat embryonic fibroblasts undergoing senescence. Ann N Y Acad Sci 851: 466–469

Guarente L (1996) Do changes in chromosomes cause aging ? Cell 86: 9–12

Guarente L (1997) Link between aging and the nucleolus. Genes Dev 11: 2449–2455

Haber DA (1997) Splicing into senescence: the curious case of p16 and p19ARF. Cell 91: 555–558

Hara E, Smith R, Parry D, Tahara, H, Stone S, Peters G (1996) Regulation of p16CDKN2 expression and its implications for cell immortalization and senescence. Mol. Cell Biol 16: 859–867

Harley CB, Vaziri H, Counter CM, Allsopp RC (1992) The telomere hypothesis of cellular aging. Exp Gerontol 27: 375–382

Hayflick L, Moorhead PS (1961) The serial cultivation of human diploid cell strains. Exp Cell Res 25: 585–621

Holliday R (1987) The inheritance of epigenetic defects. Science 238: 163–170

Holliday R (1995) Understanding Ageing. Cambridge University Press Cambridge

Iatropoulos MJ, Williams GM (1996) Proliferation markers. Exp Toxicol Pathol 48: 175–181

Jost CA, Marin MC, Kaelin WG Jr (1997) p73 is a human p53-related protein that can induce apoptosis. Nature 389: 191–194

Kaghad M, Bonnet H, Yang A, Creancier L, Biscan JC, Valent A, Minty A, Chalon P, Lelias JM, Dumont X, Ferrara P, McKeon F, Caput D (1997) Monoallelically expressed gene related to p53 at 1p36, a region frequently deleted in neuroblastoma and other human cancers. Cell 90: 809–819

Kaneko S, Satoh Y, Ikemura K, Konishi T, Ohji T, Karasaki Y, Higashi K, Gotoh S (1995) Alterations of expression of the cytoskeleton after immortalization of human fibroblasts. Cell Struct Funct 20: 107–115

Kaul SC, Wadhwa R, Komatsu Y, Sugimoto Y, Mitsui Y (1993) On the cytosolic and perinuclear mortalin: an insight by heat shock. Biochem Biophys Res Commun 193: 348–355

Kaul SC, Mitsui Y, Komatsu Y, Reddel RR, Wadhwa R. (1996) A highly expressed 81 kDa protein in immortalized mouse fibroblast: its proliferative function and identity with ezrin. Oncogene 13: 1231–1237

Kaul SC, Matsui M, Takano S, Sugihara T, Mitsui Y, Wadhwa R (1997) Expression analysis of mortalin, a unique member of the Hsp70 family of proteins, in rat tissues. Exp Cell Res 232: 56–63

Kaul SC, Mitsui Y, Wadhwa R (1998a) Molecular insights to cellular mortality and immortalization. Ind J Exp Biol 36: 345–352

Kaul SC, Duncan EL, Englezou A, Takano S, Reddel RR, Mitsui Y, Wadhwa R (1998b) Malignant transformation of NIH 3t3 cells by overexpression of mot-2 protein. Oncogene 17: 907–911

Kim NW (1997) Clinical implications of telomerase in cancer. Eur J Cancer 33: 781–786

Kubbutat MH, Jones SN, Vousden KH (1997) Regulation of p53 stability by Mdm2. Nature 387: 299–303

Kulju KS, Lehman JM (1995) Increased p53 protein associated with aging in human diploid fibroblasts. Exp Cell Res 217: 336–345

Lin AW, Barradas M, Stone JC, van Aelst L, Serrano M, Lowe SW (1998) Premature senescence involving p53 and p16 is activated in response to constitutive MEK/MAPK mitogenic signaling. Genes Dev 12: 3008–3019

Massa SM, Longo FM, Zuo J, Wang S, Chen J, Sharp FR (1995) Cloning of rat grp75, an hsp70-family member, and its expression in normal and ischemic brain. J Neurosci Res 40: 807–819

Merrick BA, Patterson RM, Witcher LL, He C, Selkirk JK (1994) Separation and sequencing of familiar and novel murine proteins using preparative two-dimensional gel electrophoresis. Electrophoresis 15: 735–745

Mizzen LA, Kabiling AN, Welch WJ (1991) The two mammalian mitochondrial stress proteins, grp 75 and hsp 58, transiently interact with newly synthesized mitochondrial proteins. Cell Regul 2: 165–179

Nakabayashi K, Ogata T, Fujii M, Tahara H, Ide T, Wadhwa R, Kaul SC, Mitsui Y, Ayusawa D. (1997) Decrease in amplified telomeric sequences and induction of senescence markers by introduction of human chromosome 7 or its segments in SUSM-1. Exp Cell Res 235: 345–353

Ning Y, Weber JL, Killary AM, Ledbetter DH, Smith JR, Pereira-Smith OM (1991) Genetic analysis of indefinite division in human cells: evidence for a cell senescence-related gene(s) on human chromosome 4. Proc Natl Acad Sci USA 88: 5635–5639

Noble JR, Rogan EM, Neumann AA, Maclean K, Bryan TM, Reddel RR (1996) Association of extended in vitro proliferative potential with loss of p16INK4 expression. Oncogene 13: 1259–1268

Noda A, Ning Y, Venable SF, Pereira-Smith OM, Smith JR (1994) Cloning of senescent cell-derived inhibitors of DNA synthesis using an expression screen. Exp Cell Res 211: 90–98

Ogata T, Ayusawa D, Namba M, Takahashi E, Oshimura M, Oishi M (1993) Chromosome 7 suppresses indefinite division of nontumorigenic immortalized human fibroblast cell lines KMST-6 and SUSM-1. Mol Cell Biol 13: 6036–6043

Ogryzko VV, Hirai TH, Russanova VR, Barbie DA, Howard BH (1996) Human fibroblast commitment to a senescence-like state in response to histone deacetylase inhibitors is cell cycle dependent. Mol Cell Biol 16: 5210–5218

Osada M, Ohba M, Kawahara C, Ishioka C, Kanamaru R, Katoh I, Ikawa Y, Nimura Y, Nakagawara A, Obinata M, Ikawa S (1998) Cloning and functional analysis of human p51, which structurally and functionally resembles p53. Nat Med 4: 839–843

Oshimura M, Barrett JC (1997) Multiple pathways to cellular senescence: role of telomerase repressors. Eur J Cancer 33: 710–715

Palmero I, Pantoja C, Serrano M (1998) p19ARF links the tumour suppressor p53 to Ras. Nature 395: 125–126

Pereira-Smith OM, Smith JR (1988) Genetic analysis of indefinite division in human cells: identification of four complementation groups. Proc Natl Acad Sci USA 85: 6042–6046

Prowse KR, Greider CW (1995) Developmental and tissue-specific regulation of mouse telomerase and telomere length. Proc Natl Acad Sci USA 92: 4818–4822

Rattan SI (1996) Cellular and molecular determinants of ageing. Indian J Exp Biol 34: 1–6

Reddel RR, Bryan TM, Murnane JP (1997) Immortalized cells with no detectable telomerase activity. A review. Biochemistry (Mosc) 62: 1254–1262

Sandhu AK, Hubbard K, Kaur GP, Jha KK. Ozer HL, Athwal RS (1994) Senescence of immortal human fibroblasts by the introduction of normal human chromosome 6. Proc Natl Acad Sci USA 91: 5498–5502

Sasaki M, Honda T, Yamada H, Wake N, Barrett JC, Oshimura M (1994) Evidence for multiple pathways to cellular senescence. Cancer Res 54: 6090–6093

Serrano M, Lin AW, McCurrach ME, Beach D, Lowe SW (1997) Oncogenic ras provokes premature cell senescence associated with accumulation of p53 and p16INK4a. Cell 88: 593–602

Seshadri T, Campisi J (1990) Repression of c-fos transcription and an altered genetic program in senescent human fibroblasts. Science 247: 205–209

Sherr CJ (1998). Tumor surveillance via the ARF-p53 pathway. Genes Dev 12: 2984–2991

Sherwood SW, Rush D, Ellsworth JL, Schimke RT (1988) Defining cellular senescence in IMR-90 cells: a flow cytometric analysis. Proc Natl Acad Sci USA 85: 9086–9090

Smith JR, Pereira-Smith OM (1996) Replicative senescence: implications for in vivo aging and tumor suppression. Science 273: 63–67

Sugawara O, Oshimura M, Koi M, Annab LA, Barrett JC.(1990) Induction of cellular senescence in immortalized cells by human chromosome 1. Science 247: 707–710

Takano S, Wadhwa R, Yoshii Y, Nose T, Kaul SC, Mitsui Y (1997) Elevated levels of mortalin expression in human brain tumors. Exp Cell Res 237: 38–45

ATrink B, Okami K, Wu L, Sriuranpong V, Jen J, Sidransky D (1998) A new human p53 homologue. Nat Med 4: 747–748

Vaziri H, Benchimol S (1996) From telomere loss to p53 induction and activation of a DNA-damage pathway at senescence: the telomere loss/DNA damage model of cell aging. Exp Gerontol 31: 295–301

von Zglinicki T, Saretzki G (1997) Molecular mechanisms of senescence in cell culture. Z. Gerontol Geriatr 30: 24–28

Wadhwa R, Kaul SC, Ikawa Y, Sugimoto Y (1991a) Protein markers for cellular mortality and immortality. Mutat Res 256: 243–254

Wadhwa R, Ikawa Y, Sugimoto Y (1991b) Natural and conditional ageing of mouse fibroblasts: genetic vs. epigenetic control. Biochem Biophys Res Commun 178: 269–275

Wadhwa R, Kaul SC, Ikawa Y, Sugimoto Y (1993a) Identification of a novel member of mouse hsp70 family. Its association with cellular mortal phenotype. J Biol Chem 268: 6615–6621

Wadhwa R, Kaul SC, Sugimoto Y, Mitsui Y (1993b) Induction of cellular senescence by transfection of cytosolic mortalin cDNA in NIH 3T3 cells. J Biol Chem 268: 22239–22242

Wadhwa R, Kaul SC, Mitsui Y, Sugimoto Y (1993c) Differential subcellular distribution of mortalin in mortal and immortal mouse and human fibroblasts. Exp Cell Res 207: 442–448

Wadhwa R, Kaul SC, Mitsui Y (1994) Cellular mortality to immortalization: mortalin. Cell Struct Funct 19: 1–10

Wadhwa R, Pereira-Smith OM, Reddel RR, Sugimoto Y, Mitsui Y, Kaul SC (1995) Correlation between complementation group for immortality and the cellular distribution of mortalin. Exp Cell Res 216: 101–106

Wadhwa R, Akiyama S, Sugihara T, Reddel RR, Mitsui Y, Kaul SC (1996) Genetic differences between the pancytosolic and perinuclear forms of murine mortalin. Exp Cell Res 226: 381–386

Wadhwa, R, Takano S, Robert M, Yoshida A, Nomura H, Reddel RR, Mitsui Y, Kaul SC (1998) Inactivation of tumor suppressor p53 by mot-2, a hsp70 family member. J Biol Chem 273: 29586–29591

Wang E (1995) Senescent human fibroblasts resist programmed cell death, and failure to suppress bcl2 is involved. Cancer Res 55: 2284–2292

Wang E, Lee MJ, Pandey S (1994) Control of fibroblast senescence and activation of programmed cell death. J Cell Biochem 54: 432–439

Wareham KA, Lyon MF, Glenister PH, Williams ED (1987) Age related reactivation of an X-linked gene. Nature 327: 725–727

Webster TJ, Naylor DJ, Hartman DJ, Hoj PB, Hoogenraad NJ (1994) cDNA cloning and efficient mitochondrial import of pre-mtHSP70 from rat liver. DNA Cell Biol. 13: 1213–1220

Weinberg RA (1996) The molecular basis of carcinogenesis: understanding the cell cycle clock. Cytokines Mol Ther 2: 105–110

Whitaker NJ, Bryan TM, Bonnefin P, Chang AC, Musgrove EA, Braithwaite AW, Reddel RR (1995) Involvement of RB-1, p53, p16INK4 and telomerase in immortalisation of human cells. Oncogene 11: 971–976

Wynford-Thomas D (1996a) p53: guardian of cellular senescence. J Pathol 180: 118–121

Wynford-Thomas D (1996b) Telomeres, p53 and cellular senescence. Oncol Res 8: 387–398

Yeager TR, Stadler W, Belair C, Puthenveettil J, Olopade O, Reznikoff C. (1995) Increased p16 levels correlate with pRb alterations in human urothelial cells. Cancer Res. 55: 493–497

Yeager TR, DeVries S, Jarrard DF, Kao C, Nakada SY, Moon TD, Bruskewitz R, Stadler WM, Meisner LF, Gilchrist KW, Newton MA, Waldman FM, Reznikoff CA (1998) Overcoming cellular senescence in human cancer pathogenesis. Genes Dev 12: 163–174

Zhang Y, Xiong Y, Yarbrough WG (1998) ARF promotes MDM2 degradation and stabilizes p53: ARF-INK4a locus deletion impairs both the Rb and p53 tumor suppression pathways. Cell 92: 725–734

Zhu J, Woods D, McMahon M, Bishop JM (1998) Senescence of human fibroblasts induced by oncogenic Raf. Genes Dev 12: 2997–3007.

Subject Index

Printing: Druckhaus Beltz, Hemsbach
Binding: Buchbinderei Schäffer, Grünstadt